NATURAL HISTORY OF THE ANTARCTIC PENINSULA

NATURAL HISTORY OF THE ANTARCTIC PENINSULA

Text by
SANFORD MOSS

Illustrations by
LUCIA deLEIRIS

COLUMBIA UNIVERSITY PRESS
NEW YORK

Columbia University Press

New York   Chichester, West Sussex

Copyright © 1988 Columbia University Press
Illustrations copyright © 1988 by Lucia deLeiris
All rights reserved

Library of Congress Cataloging-in-Publication Data

Moss, Sanford A.
  Natural history of the Antarctic Peninsula.

  Includes bibliographies and index.
  1. Natural history—Antarctic regions—Antarctic
Peninsula.   I. deLeiris, Lucia.   II. Title.
QH84.2.M67   1988   508.98'9   87-10971
ISBN 0-231-06268-0
ISBN 0-231-06269-9 (pbk.)

Clothbound editions of Columbia University press are
printed on permanent and durable acid-free paper.

Printed in the United States of America

Book design by Ken Venezio

c 10 9 8 7 6 5 4 3 2 1
p 10 9 8 7 6 5

CONTENTS

PREFACE

Antarctica is a continent of superlatives. It is the coldest, the driest, the windiest, the iciest, and (with its ice cap) averages as the highest in altitude of all the major land masses of our world. It is the continent with the longest nights, the longest days, the least amount of soil, the greatest amount of fresh water (all locked up as ice), and it is surrounded by the stormiest ocean on earth—the Southern Ocean.

Antarctica is also a continent of contrasts. While having 90 percent of the earth's ice and snow it is also the driest continent in terms of relative humidity and annual precipitation—less than 15 centimeters (6 inches) of rain equivalent a year. Antarctica has a rich history of exploration and discovery by adventurers representing an array of nations, yet it presents a showcase of international cooperation where, by treaty, national territorial claims are held in abeyance. This is the most remote and environmentally hostile of all the continents, but its polar position and relative isolation from man's despoilations make it the most valuable window we have to events going on in our outer atmosphere and the far reaches of space. Because its ice and snow are locked in perpetuity, the depths of the Antarctic ice sheet (which averages about 2,160 meters deep) have deposited in their accumulated layers a record of the earth's atmospheric composition that reaches back tens of thousands of years. So this—the most remote and unfriendly continent—is, scientifically speaking, our most valuable.

A most important contrast concerns the principal focus of this book. Even though Antarctica is the fifth largest of the continents, it has the fewest forms of life inhabiting it. This fact provides unparalleled opportunities for naturalists. The plants and animals that visit, breed, and in some instances thrive here are of special interest to students of natural history. They offer one of the least complex webs of ecological interrelationships to be found on earth. This is a place for the ecologist to formulate and test theory. At the same time the adaptations which make life possible for antarctic plants and animals are often very special indeed. They make wonderful study opportunities for the anatomist, the physiologist, and the evolutionist.

The rest of us can also marvel at the living spectacles that Antarctica places before us. There is beauty here, both on the grand scale of an enormous blue whale, or of a breeding colony of tens of thousands of Adélie Penguins, as well as on the microscopic scale of a tiny mite clinging tenaciously to a life dependent on the slowly growing rhizomes of a century-old moss plant. It can do all of us some good to understand the curious molar teeth of the crabeater seal, or the burrowing nesting habit of Wilson's Storm Petrel—both examples of the sorts of adaptations which permit life to exist on this austere continent.

Although the great expanses of the central plateau and inland mountain ranges of Antarctica are virtually lifeless, the seas that fringe the continent moderate its harsh climate enough so that an array of life forms can be found near the coast. Nowhere are these organisms more plentiful or diverse than on the Antarctic Peninsula, the twisted panhandle that stretches north toward the grasping reach of Tierra del Fuego at the extreme southern end of South America. Here we find the living stuff for this book.

Most works on antarctic life deal with the obvious—the penguins, the seals, the whales. This book too treats these animals, but especially from the points of view of their unique adaptations and their ecological relationships. This latter perspective will take us to the smaller and more cryptic plants and animals of the Antarctic--organisms that are less well known, but that are just as important as the frequently photographed seals, seabirds, and penguins.

It is difficult to write of living organisms without considering the physical conditions that form the backdrops against which they play out their lives. To that purpose this book begins by describing the physical conditions—the geology, geography, and climate of the antarctic continent, and the oceanography of the Southern Ocean that surrounds it. Then to provide the most coherent account, we will trace the flow of biological energy through the terrestrial and oceanic ecosystems of the Antarctic. Beginning with the primary producers—the plants that "eat" sunlight—we will follow their energetic substance as it gets passed from body to mouth through all the trophic levels of these ecosystems. This approach, we hope, will give you, the reader, a richer fuller appreciation of the conditions of antarctic life, its interrelationships, and the special adaptations that make it possible. With this understanding there is a form and substance to the antarctic existence that we find exhilarating and inspiring. We hope that you will too.

To provide a perspective unique among books about Antarctica, this one features the illustrations of artist Lucia deLeiris who spent four months at Palmer Station on the Antarctic Peninsula, and on board ship along its length, creating the drawings reproduced here. In this way the organisms—microscopic and cryptic, small or gigantic—can be presented with a perspective, form, and feeling that are impossible to achieve photographically. Many of the illustrations are captioned with excerpts from the journals kept by Lucia while she was in the Antarctic. We hope these impressions will help to convey a deeper understanding and appreciation of this exceptional continent.

Lucia's visit to the Antarctic was supported by the National Science Foundation's Division of Polar Programs. We are greatly indebted to this organization and to its staff, particularly Guy Guthridge for the help and courtesy that he provided. Don Wiggin, manager of Palmer Station, extended a great deal of help and facilitated Lucia's work. We also thank Monte Snyder and the crews of the *Polar Duke* and the *Arctic Survey Boat* for their assistance.

A great many other people gave freely of their time and expertise to make this book possible. The following scientists critically reviewed drafts of chapters concerning their specialties, and offered suggestions that materially improved the writing: William Detrich, Department of Biochemistry, University of Mississippi Medical Center, Jackson, Mississippi; Langdon Quetin and Robin Ross, Marine Science Institute, University of California, Santa Barbara; Brian Obst and George Matsumoto, Biology Department, University of California at Los Angeles; Rob Willan, British Antarctic Survey; and Warren Zapol, Department of Anesthesiology, Massachusetts General Hospital. Other scientists who provided useful information include: Carl Boyd, Department of Oceanography, Dalhousie University, Halifax, Nova Scotia; Robert Burger, Geology Department, Smith College; Robert Edgar, Biology Department, Southeastern Massachusetts University; Peggy Hamner, Biology Department, University of California, Los Angeles; Andrzej Myrcha, Institute of Ecology, Polish Academy of Sciences; and J. Treadwell Turner, Biology Department, Southeastern Massachusetts University. While all of these individuals provided assistance, any errors in the book are the sole responsibility of the author.

We also wish to acknowledge the following individuals and organizations for their assistance in producing this book: Barbara Moss, Sheila Freitas, the library staff at Southeastern Massachusetts University, the Farlow Herbarium at Harvard University, and the editorial staff at Columbia University Press, including Susan Koscielniak, Edward Lugenbeel, and Donna Walsh. Without their help our task would have been more difficult.

CHAPTER 1

THE LANDSCAPE

Antarctica, the most recently discovered and fifth largest continent, lies nearly centered on the earth's southern rotational pole. Practically all of the continent, save small parts of the Antarctic Peninsula and East Antarctica, are within the Antarctic Circle at 66° 30' South latitude. Thus, in the austral summer, from mid-December to mid-March, the land enjoys sunlight twenty-four hours a day; and in the southern winter it suffers no glimpse of the sun whatsoever.

The circular symmetry of Antarctica is marred by two indentations in the land. The eastern one (below the Atlantic Ocean) is the Weddell Sea, penetrating toward the south pole between longitudes 30° and 60° West, and under its floating ice sheet to about 82° South latitude. The Ross Sea lies on the western shoulder of the continent between longitudes 170° and 150° East, and reaches under its floating ice sheet to about 85° South latitude. Between these massive embayments the rugged Antarctic Peninsula, like the tail of the ray-shaped continent, projects from its base in West Antarctica to the northeast, between longitudes 55° and 75° West, terminating in a cluster of rugged islands at about 62° South.

THE ICE SHEET

Antarctica covers an area of almost 14 million square kilometers (almost 5,400,000 square miles), although uncertainties about the actual extent of embayments under ice that is fast to the shore make this figure approximate. Probably less than 3 percent of the land area of Antarctica is free of ice and snow at least part of the year, the rest residing under a layer of glacial ice averaging 2,160 meters thick with a maximum depth of about 4,000 meters. The volume of this ice sheet is immense. It contains 2 percent of all of the water on earth, and 80 to 90 percent of its fresh water. The ice sheet's crushing weight has depressed the antarctic land mass under it, and, were it to melt, its water could raise the sea level of all our oceans about 60 meters—enough to drown the principal coastal cities of the world, as well as many inland ones. Although there is not much doubt that this ice sheet will someday melt, that spectacle will probably not happen quickly.

The glacial ice that buries the continent originates as snow packed into ever denser annular layers, which, as they accumulate, are increasingly compressed by the gathering weight of the younger layers above them. Gradually the air in the packed snow is expressed and, under greater pressure, the ice granules become fused into glacial ice. Although the ice sheet began to originate as early as Miocene time, about 25 million years ago, and covered the continent 15 million years ago, probably no more than about 700,000 years' accumulation of ice exists at the present time.

The antarctic ice cover produces phenomena that profoundly affect not just this continent, but physical and biological conditions throughout the

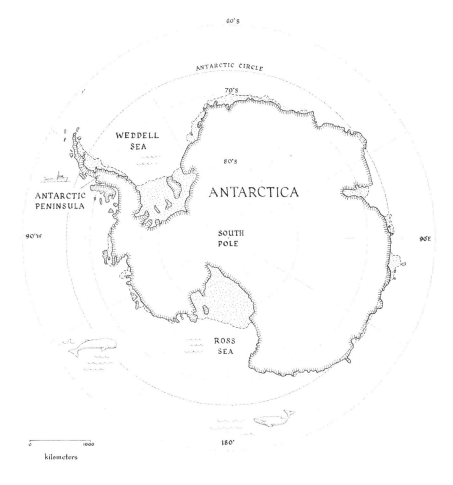

world. In addition to the huge reservoir of water that it contains, the ice sheet has vast climatological importance. As an immense site of permanent ice and snow cover, Antarctica is a major factor in determining the reflectivity (albedo) of the Southern Hemisphere. When sunlight reflects off the snow, glacial ice, and sea ice of the Antarctic, it radiates back to space without being transformed into heat, and is lost as a source of energy to drive large scale processes of oceanic and atmospheric circulation. Increasing the albedo by extending the permanent ice cover of the Southern Ocean, or its decrease by ice cap melting and ablation would have

Section through the continent

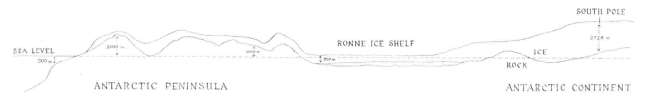

Glacier edge

profound effects on the climate of the Southern Hemisphere. In the former case the positive feedback of greater reflectivity could produce ever colder average temperatures.

The ablation of the ice sheet (reducing the albedo) could raise temperatures and signal a new tropical era for those parts of the earth that escape drowning. That both of these scenarios have occurred more than once over only the past one million years of the earth's history is testimony to the cyclic nature of climatic processes that affect our planet. Indeed, Antarctica is sometimes considered to be the "author of global climate."

The ice sheet, with its great weight averaging well over a ton per square inch, plays several geologically important roles in the Antarctic. First, its burden depressed the continent into the underlying mantle of the earth, and today significant portions of land surface are well below sea level. The Antarctic Peninsula is connected through such a low-lying trough under the ice sheet to the rest of the continent. To geologists Antarctica is really two continents, East Antarctica and West Antarctica (the Peninsula and Marie Byrd Land). Others argue that the Peninsula represents an offshore island mass connected to the mainland by the ice sheet.

When the ice sheet ultimately melts, and the pressure on the continent is relieved, the crust will slowly rebound and rise to a new altitude. The magnitude of this isostatic movement can be estimated by looking at the continental shelf. The shelves of continents unburdened by ice exist at depths under the sea surface of about 200 meters. The antarctic continental shelf, however, lies 500 to 600 meters deep—and may extend, in portions of East Antarctica to 1,000 meters. The eventual rebound, when melting takes place, will be perhaps 300 to 400 meters.

The ice sheet flows outward to the Southern Ocean in all directions. When in equilibrium it loses by ablation at its outer edges as much ice as snow accumulates on its surface each year. Some of this loss occurs as evaporation (sublimation) of water vapor directly from the ice surface, but much is due to the calving of icebergs from the ice edge. Where the ice sheet projects over ocean embayments (as in the Ross Sea, the Weddell Sea, and elsewhere) it can fracture into immense tabular bergs that measure as much as hundreds of square kilometers in surface area. Where the ice funnels to the coast in glacial streams, irregular blocks break off and fall into the sea with thunderous roars that can be heard for miles.

A significant consequence of the ice sheet is that it obscures direct examination of the rocks that compose the crust of this part of the earth. Only the isolated rock pinnacles (called nunataks) that project above the ice and snow, the most precipitous slopes of the mountainous regions, and the relatively ice-free islands of the Antarctic Peninsula are available for geologists to study directly. The most accessible and extensive ice-free areas include the so-called "dry valleys," the best studied of which lie in Victoria Land, not far from the U.S. research station at McMurdo Sound. These valleys are unobscured by snow or ice year-round and present a veritable mine of information about the geology of the Antarctic.

A tabular iceberg

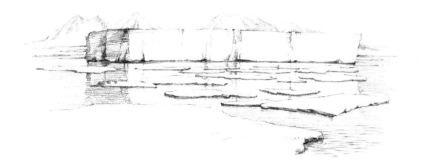

GEOLOGY

Despite the overwhelming cover of ice and snow in Antarctica, geologists have pieced together a reasonably good picture of crustal composition and the geological history of this austere continent.

Two huge units of crust (continental tectonic plates in the jargon of geology) make up Antarctica. The largest of these is the vast east antarctic plate or craton that is nearly completely covered by the main antarctic ice sheet. The second unit is complex and makes up the Antarctic Peninsula and West Antarctica west of a line drawn from McMurdo Sound on the Ross Sea through to the eastern edge of the Weddell Sea. This is the landscape central to this book. These units ride on the surface of the mantle—a 2,900 kilometer thick layer of rock that surrounds the core of the earth. The core itself is 6,900 kilometers thick and is composed of dense iron-rich material that is consolidated into a solid center and a fluid outer layer. It is thought that the motion of the fluid layer of the core produces the earth's magnetic field, the electrically negative pole of which projects off the coast of East Antarctica as the south magnetic pole.

Relationship of magnetic pole to rotational south pole

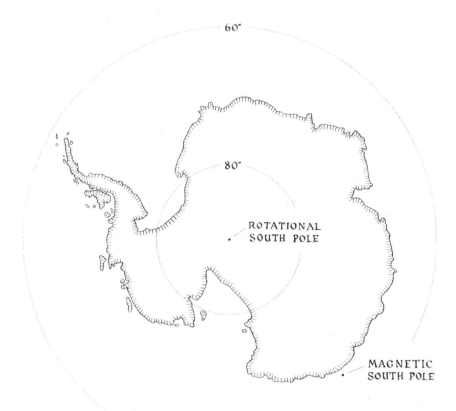

The magnetic poles, both north and south, are not coincident with the north-south axis of rotation. Moreover, the magnetic poles reverse polarity periodically (about every half-million years or so) and affect the orientation of magnetically susceptible minerals when they crystallize into new rock, permanently recording the pole position at the time the rock is formed. The most recent pole reversal occurred about 700,000 years ago. Indeed, the magnitude of the earth's magnetic field has declined about 6 percent in just the last 150 years. Perhaps this presages a pole reversal—an event sure to have profound consequences for compass navigators, electronic communications, and animals like birds, sharks, and honeybees that can sense and orient to the earth's magnetic field.

The geologic structure of West Antarctica differs significantly from that of the eastern end of the continent. Here the basement layer of the crust lies buried 2 or 3 kilometers deep under an accumulation of sedimentary rock, and the crust itself is only 25 or 30 kilometers thick. The Transantarctic Mountains, which run eastward from Victoria Land to the southern rim of the Weddell Sea mark an abrupt change from the thin surface layer and thick crust that characterize East Antarctica to the thick surface layer and thin crust of West Antarctica. Just to the north of these mountains a belt of subsidence occurs, where, under the ice sheet, deep troughs—as much as 2,000 meters below sea level—exist. This depression represents a boundary between the edges of at least two units of the earth's crust.

The highest mountains in Antarctica—the Vinson Massif at more than 5,100 meters (about 16,750 feet) in the Ellsworth Mountains—lie just to the north of this boundary, and could represent a broken piece of the East Antarctic shield. The remainder of West Antarctica, including the Peninsula, contains a hodge-podge of at least three, and perhaps more, small plates. Volcanism occurs over the length of the Antarctic Peninsula, and has been profound in the past.

The Antarctic Peninsula has a complex system of sedimentary basins, and because more of it is exposed to direct geological examinations, it has a better understood history. The geologically recent sedimentary and volcanic activity here resulted in the entombment and fossilization of a host of plants and animals; and in places, the formation of some coal. Fossil remains here date from the Cambrian Period of the Paleozoic Era, which began approximately 600 million years ago. They include primitive sponges, trilobites, and algae found in limestones in the Ross Archipelago of the Peninsula.

From the Mesozoic Era are found spectacularly rich and luxuriant plant beds at Mt. Flora on Hope Bay that date from the Jurassic, about 160 million years ago. More than 60 kinds of plants have been described from these strata, along with freshwater fish and beetle remains. Other Jurassic fossils include bivalves, brachiopods, and crinoids collected from Ellsworth Land at the base of the Peninsula.

''. . . strange contorted shapes of blue ice, distant mountains with steep edges and sharp peaks. These mountains are spectacular. Some peaks are fog-shrouded and have a mystical look. Where the sun breaks through the overcast it sometimes lights up an iceberg against a misty blue background with such brilliance that it shines as though lit from within.''

Still later in the Mesozoic are abundant Cretaceous fossil beds (120 to 70 million years old) on the east side of the Peninsula and through the James Ross Archipelago that include marine-derived flora and fauna, particularly rich in foraminifera, ammonites, and various bivalve mollusks.

There are but few fossil mesozoic vertebrates in Antarctica. These include the scattered remains of mosasaurs and plesiosaurs (marine reptiles) dating from the late Cretaceous on Seymour Island.

Cretaceous fossils from this region are historically interesting because they were observed and commented on by the famous pioneering antarctic whaler, Captain C. A. Larsen, after his first visit to the Peninsula in 1891. Larsen's observations resulted in his guiding the 1901–1903 Swedish Antarctic Expedition, led by Otto Nordenskjöld, to the Snow Hill—

Seymour Island region of the Peninsula. Larsen's ship, the *Antarctic,* was lost in the Weddell ice pack and Nordenskjöld's party was forced to spend a second, and unexpected, winter in the Antarctic. The geological investigations carried out by this expedition were published by Andersson in 1906, and remain one of the finest geological descriptions of the Antarctic Peninsula.

Seymour Island proved to be a veritable gold mine of fossil remains dating from the end of the Cretaceous (70 million years ago) into the upper Eocene of the Tertiary, about 40 million years ago. The Paleocene and Eocene deposits are dominated by mollusks, but virtually every important group of invertebrates (sans sponges) are found here, and some of these are the earliest representatives of their groups known in the fossil record anywhere.

Seymour Island Tertiary vertebrate remains include shark teeth and fossil

Fossil fern

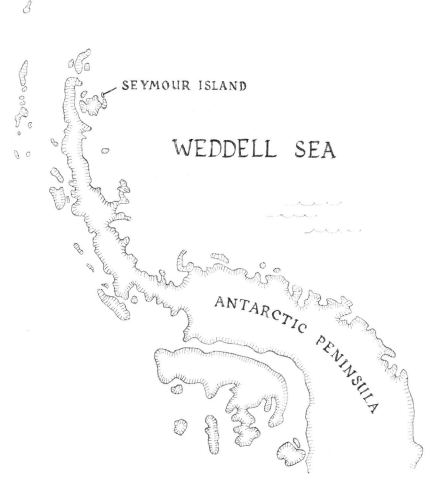

SEYMOUR ISLAND

WEDDELL SEA

ANTARCTIC PENINSULA

Seymour Island

Anarctolops jawbone

penguins, but the most important find was the first mammal *(Anarctolops)* found anywhere in Antarctica. The remains of this squirrel-sized animal consist of fragments of the upper and lower jaws (with teeth) and establish it as a marsupial, related to others known from the same time in North and South America. This fossil has important implications for the existence of land connections between South America, Antarctica, and Australia—connections that will be discussed below.

In many late Mesozoic and early Cenozoic deposits in Antarctica, woody plants are known, including petrified trees as thick as half a meter in diameter. Other interesting fossils are footprints and trackways of birds. For example, the footprints of large ground birds as well as large ducks or geese are known from Oligocene and Miocene deposits (20 to 30 million years ago) on King George Island.

Pleistocene to Recent fossil remains in the Antarctic Peninsula are dominated by shell heaps (apparently concentrated by birds or by wave and current action—not by humans) that contain several genera of bivalves, including the scallop, *Pecten*. Subfossil remains of penguins are locally abundant at sites of prior rookeries.

Given the small percentage of snow- and ice-free land surface, and the hostile weather conditions that prevail, there is a rich assemblage of fossils here. Much of the fossil flora and fauna are representative of climates more equable than that of contemporary Antarctica. This observation, along with parallel examples of subtropical flora and fauna in high northern latitudes (Alaska and Spitzbergen are but two examples) suggest both that the earth was previously warmer than at present, and that the relative positions of the continents have changed over life's fossil history. There is plenty of evidence to support the first hypothesis, and many, if not most, earth scientists consider the latter possibility to be true.

CONTINENTAL DRIFT

As early as 1912 a German geologist, Alfred Wegener, summarized the work of earlier geologists and proposed that similarities in geological structure, complementary continental outlines, and related fossil assemblages suggest a former association of land masses much different than those that we see today. He hypothesized a world of drifting continents; land masses slowly moving independently in an inexorable way.

Wegener's model was fifty years ahead of its time, for most of his contemporary geologists and biogeographers in the Northern Hemisphere found this incredible thesis untenable. By the early 1960s, however, a wealth of geophysical information had accumulated that pointed to the conclusion reached by Wegener. These data included paleomagnetic information that earlier had been interpreted as due to "wandering" magnetic poles—not movement of land masses—and observations that even the deepest and oldest ocean sediments are geologically young. Moreover, it was learned that the youngest bedrock on the oceans' bottoms are found in the system of mid-ocean ridges that lace the great ocean basins of the world. These ridges were found to be the sources of new sea floor.

A modern reading of the geophysical data related to continental drift has a southern supercontinent, Gondwana, splitting by the middle Jurassic (about 160 million years ago) into two great blocks. The eastern one ultimately spawned Australia, Antarctica, India, and Madagascar. The western block fractured into South America and Africa. By 115 million years ago, in the early Cretaceous, the spreading continental masses had moved southward (most of the Peninsula and West Antarctica was between 50° and 60° South latitude), and South America and Africa had moved apart, creating the South Atlantic Ocean between them. What is today the northern tip of the Antarctic Peninsula (then trending east to west) lay adjacent to the tip of South America. It maintained that position until at least the middle of the Tertiary (35 million years ago).

Even though paleomagnetic data point to a far southern location of the Antarctic Peninsula at this time, analyses of the fossil flora present in the early Cretaceous (100 to 120 million years ago) suggest that the Mesozoic climate of the earth may have been quite different than the one we experience today. Fossil leaves and tree trunks found on Alexander Island

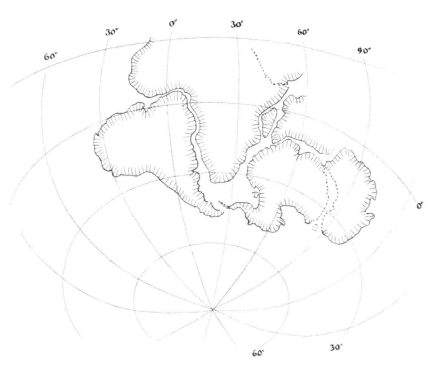

The breakup of Gondwana, mid-Jurassic. Redrawn and modified from Norton, 1982

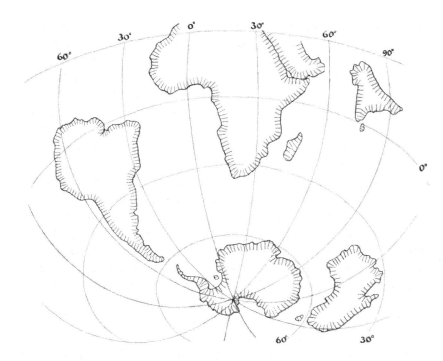

Peninsula-South America relationship 40 million years ago

near the base of the Peninsula show that these plants grew under conditions similar to those of warm temperate rain forests today. The paleomagnetic data point to a location of Alexander Island at 70° to 80° South at this time. The Drake Passage of today did not exist. The proximity of the Antarctic Peninsula to South America provided the route taken by marsupials in their movement from South America to Australia. It was only in the early Eocene (53 million years ago) that Australia began to separate from East Antarctica with a mid-ocean ridge developing as sea floor spreading accommodated that movement.

During the Oligocene epoch (38 to 29 million years ago) the final southern rotation of Antarctica occurred, and, most significantly, the Drake Passage grew as a deepwater channel. This finally allowed the present circumpolar current to develop, and with it the climatic isolation that so clearly defines present-day Antarctica. The history of the modern antarctic ice sheet thus probably does not predate the late Oligocene (*ca* 28 million years ago). The ice sheet was probably not completely formed until only 14 or 15 million years ago.

Since the Oligocene Antarctica has stabilized and its present position is essentially that which it occupied 30 million years ago. Despite active volcanism along the boundaries of its micro plates, the Antarctic Peninsula today is seismically quiet. Earthquakes typical of tectonic plate movement are rare.

CLIMATE

The major features of climate include temperature, wind, precipitation, and solar radiation. Among the continents of the world, Antarctica presents extremes in all of these climatic categories. It is the coldest, windiest, driest continent, and also, because of its albedo, the one with the lowest absorbed solar radiation. All of these superlatives relate to the fact that the continent is polar in position; permanently snow-covered; dome-shaped with a high central plateau that drops to the coast, and is surrounded by a warmer, energy-rich ocean.

Temperature

Antarctica is cold. Air and surface temperatures through much of the interior never rise above the freezing point (0°C), and for most of the year are well below this. The annual mean temperature in the coldest central region is below −55°C. The coldest temperature ever measured at the earth's surface, −89.6°C, was observed at the Soviet's Vostok Station on the central plateau. At the U.S. South Pole Station the seasonal mean temperatures are: winter, −59°C; spring, −49°C; summer −32°C; and fall, −57°C. At the coast the temperatures become more equable with typical seasonal means on the Antarctic Peninsula being: winter, −8 to −20°C; spring, −3 to −11° C; summer, +1 to −2°C; and fall, −2 to −14°C.

The difference between interior and coastal temperatures reflects a number of factors. First, air will lose 1°C for each 100 meters of altitude gained. The South Pole lies at an elevation of 2,800 meters. Second, the percentage of cloud cover is much greater in coastal than in interior regions. This means that radiational cooling to the open sky is more prominent on the central plateau. When radiational cooling occurs it produces temperature inversions (the coldest layers are nearest the ground). These inversions on the coast are more likely to be broken up by winds (the coast is windier) which mix warmer air from altitude, raising the surface temperature. A most important factor is the presence of the ocean. With water temperatures only a degree or two below zero, and sea ice in virtual thermal equilibrium with the water, a tremendous source of heat is available to be transferred to the atmosphere. Finally, the antarctic coastline, and especially the Peninsula, has most of the bare ground found in the continent. Here sunlight can be absorbed as heat, rather than reflected back into space. Heat absorption budget surpluses of more than 37,000 calories per square centimeter per year have been measured in such places, making the soil surface warmer than the air for six months of the year.

Under these conditions soils can thaw (in some places to depths of 2 meters) and conditions become tenable for plant life (chapter 3).

On a monthly basis the Antarctic Peninsula shows temperatures that average warmest in January (a degree or two above zero) and coldest in June (−15°C to −20°C). The Weddell Sea side of the Peninsula is colder than on the west, a fact that is reflected in the persistence of ice shelves (Ronne, Filchner) that cling to the eastern side. These temperatures correspond with the maximum and minimum solar radiation (24 hours of sunlight per day in much of December, and no sunlight in June) south of the Antarctic Circle. Even though the minimum temperatures on the Peninsula are less extreme than those of the central plateau, the climate of the coastal regions is made harsh by the winds that regularly rake them. There are three general sources of these winds, and when two or more conspire to act together, fearsome velocities occur.

Wind

The dome shape of Antarctica gives rise to air movements that are known as "katabatic" or "gravity" winds. Cold, dense air produced by radiational cooling of the interior flows downhill. When this air slides off the edges of the central plateau and falls to the coast it gathers momentum, attaining velocities greater than 22 meters per second (50 miles per hour). The ordinary katabatic wind averages 14 meters per second. These winds are most intense at the coast in East Antarctica where the central plateau falls most steeply to the coast. They are less important on the Peninsula. Pure katabatic winds originate 20 to 25 kilometers inland and project 10 to 30 kilometers out to sea.

Katabatic winds have several interesting properties and effects. Normally they involve only the lower few hundred meters of the atmosphere and are relatively steady. Intense katabatic winds, however, pass a critical point where they become turbulent and gusty. It is then that they pick up loose snow and blow considerable quantities of it down to the coast where it is transported and deposited as far as 10 to 25 kilometers out to sea. As these winds descend they are heated adiabatically 1°C for each 100 meters of descent. This means the winds are relatively warm (although the wind chill factor keeps people from appreciating it). With the temperature rise there is a concommitent reduction in relative humidity so that this air becomes exceedingly dry.

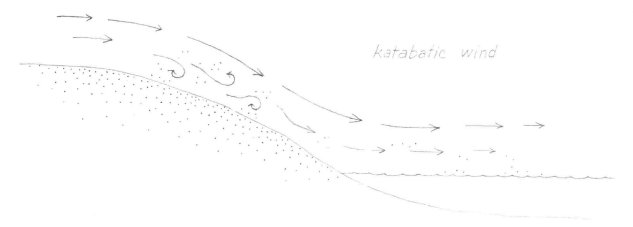

katabatic wind

A second antarctic wind component is developed by massive convection cells associated with the Southern Ocean surrounding the continent. The warmed surface air rises over the ocean and is replaced by colder air that descends from the continent, producing outward, north-blowing winds. Because the temperature differential between sea surface temperature and continental temperature is greatest during the winter, the strongest winds occur at that season of the year. On the Peninsula, the seasonal and annual mean wind velocities range between 4 and 9 meters per second (9 and 20 miles per hour). At some coastal locations in East Antarctica, fall and winter winds average 11 to 12 meters per second (25 miles per hour). Maximum wind velocities of over 100 miles per hour are regularly experienced in coastal East Antarctica, where hurricane-force winds are to be expected more than seven times in a typical winter season. Those maximum wind velocities are associated with storm systems that form over the Southern Ocean and provide the third and most intense source of antarctic winds. Altogether the wind regime of this continent is the most severe of any other, and, coupled with the extreme low temperature and low humidity, presents an exceedingly hostile environment for all terrestrial life forms.

''. . . stinging spray everywhere. The katabatic wind sliding off the glacier picked up from less than 10 knots to 52 knots in minutes, creating a huge swell. I've never seen such fast change in the weather!''

Precipitation

If Antarctica is the coldest, windiest continent, it is also the driest. Precipitation, virtually all of it snow (summer rains occur in the northern half of the Peninsula), is especially reduced in the interior where the outward-flowing winds allow little transport of moisture in from the sea. The constant blowing of snow (which creates distinctive drifts known as sastrugi) makes accurate measurement of snowfall difficult, but much of the interior accumulates less than 5 centimeters (2 inches) of water equivilent per year. As the coastline is approached, annual accumulation reaches 30 centimeters (16 inches) of water equivilency. The annual net accumulation for the whole continent has been estimated as 14.5 centimeters (5.7 inches) of water equivalent. If snow drifting, sublimation, and surface melting are taken into consideration, the average annual net precipitation is probably between 14.6 and 19.2 centimeters (5.7 and 7.6 inches. The extended blizzards so typical of this continent conceal its truly dry nature. Most of the snow moved in such events is blowing old snow, not new accumulation. Antarctica is truly a white desert.

The severe climate of Antarctica is certainly in part due to its polar position, but this climate also reflects the relationship of the continent with its encircling ocean. Physical, chemical, and biological processes occurring in that ocean profoundly affect the continent, and indeed, much of the Southern Hemisphere. The oceanography of the Southern Ocean will make an appropriate subject for the next chapter.

ADDITIONAL READING

Adie, R. J. 1970. Past environments and climates of Antarctica. In M. W. Holdgate, ed., *Antarctic Ecology*, 1:7–14. London and New York: Academic Press.

Adie, R. J. 1972 Recent advances in the geology of the Antarctic Peninsula. In R. J. Adie, ed., *Antarctic Geology and Geophysics*, pp. 121–124. Oslo: Universitetsforlaget.

Andersson, J. G. 1906. On the geology of Graham Land. *Bulletin of the Geological Institute of Upsala*, 7:19–71.

Birkenmajer, K. 1985. Onset of Tertiary continental glaciation in the Antarctic Peninsula sector (West Antarctica). *Acta Geologica Polonica*, 35(1–2):1–31.

Colbert, E. H. 1982. Mesozoic vertebrates of Antarctica. In C. Craddock, ed., *Antarctic Geoscience*, pp. 619–627. Madison: University of Wisconsin Press.

Craddock, C. 1982. Antarctica and Gondwanaland. In C. Craddock, ed., *Antarctic Geoscience*, pp. 3–14. Madison: University of Wisconsin Press.

Dalziel, I. W. D., M. J. de Wit, and C. R. Stern. 1975. Structural and petrologic studies in the Scotia Arc. *Antarctic Journal of the United States*, 10:180–186.

Farquharson, G. W. 1983. Evolution of late mesozoic sedimentary basins in the northern Antarctic Peninsula. In R. I. Oliver, P. R. James, and J. B. Jago, eds., *Antarctic Earth Science*, pp. 323–327. Cambridge: Cambridge University Press.

Farquharson, G. W. 1984. Late mesozoic, non-marine conglomeratic sequences of northern Antarctic Peninsula (the Botany Bay group). *British Antarctic Survey Bulletin*, 65:1–32.

Gonzalez-Ferran, O. 1983. The seal nunataks: An active volcanic group on the Larsen Ice Shelf, West Antarctica. In R. I. Oliver, P. R. James, and J. B. Jago, eds., *Antarctic Earth Science*, pp. 334–337. Cambridge: Cambridge University Press.

Grindley, G. W. and F. J. Davey. 1982. The reconstruction of New Zealand, Australia, and Antarctica. In C. Craddock, ed., *Antarctic Geoscience*, pp. 15–26. Madison: University of Wisconsin Press.

Hollin, J. T. 1970. Antarctic glaciology, glacial history, and ecology. In M. W. Holdgate, ed., *Antarctic Ecology*, 1:15–30. London and New York: Academic Press.

Jefferson, T. H. 1983. Palaeoclimatic significance of some mesozoic antarctic fossil floras. In R. I. Oliver, P. R. James, and J. B. Jago, eds., *Antarctic Earth Science,* pp. 593–598. Cambridge: Cambridge University Press.

Johnson, G. L., J. R. Vanney, and D. Hayes. 1982. The antarctic continental shelf. In C. Craddock, ed., *Antarctic Geoscience,* pp. 995–1,002. Madison: University of Wisconsin Press.

Norton, I. O. 1982. Paleomotion between Africa, South America, and Antarctica, and implications for the Antarctic Peninsula. In C. Craddock, ed., *Antarctic Geoscience,* pp. 99–107. Madison: University of Wisconsin Press.

Oerlemans, J. and C. J. van der Veen. 1984. *Ice Sheets and Climate.* Dordrecht: D. Reidel.

Quilty, P. G. 1982. Tectonic and other implications of middle-upper Jurassic rocks and marine faunas from Ellsworth Land, Antarctica. In C. Craddock, ed., *Antarctic Geoscience,* pp. 669–678. Madison: University of Wisconsin Press.

Rich, P. V. 1975. Antarctic dispersal routes, wandering continents, and the origin of Australia's non-passeriform avifauna. *Memoirs of the National Museum, Victoria,* 36:63–126.

Schwerdtfeger, W. 1984. *Weather and Climate of the Antarctic.* Developments in Atmospheric Science Series, vol. 15. New York; Elsevier.

Simpson, G. G. 1978. Early mammals in South America: Fact, controversy, and mystery. *Proceedings of the American Philosophical Society,* 122(5):318–328.

Thomson, M. R. A., R. J. Pankhurst, and P. D. Clarkson. 1983. The Antarctic Peninsula—a late mesozoic-cenozoic arc (review). In R. I. Oliver, P. R. James, and J. B. Jago, eds, *Antarctic Earth Science,* pp. 289–294. Cambridge: Cambridge University Press.

Van Mieghem, J. and P. Van Oye, eds. 1965. *Biogeography and Ecology in Antarctica.* The Hague: Junk.

Walton, D. W. H. 1984. The terrestrial environment. In R. M. Laws, ed., *Antarctic Ecology,* 1:1–60. London and Orlando, Fl.: Academic Press.

Woodburne, M. O. and W. J. Zinsmeister. 1982. Fossil land mammal from Antarctica. *Science,* 218:284–286.

Woodburne, M. O. and W. J. Zinsmeister. 1983. A new marsupial from Seymour Island, Antarctic Peninsula. In R. I. Oliver, P. R. James, and J. B. Jago, eds., *Antarctic Earth Science.* pp. 320–322. Cambridge: Cambridge University Press.

Zinsmeister, W. J. 1984. Geology and paleontology of Seymour Island, Antarctic Peninsula. *Antarctic Journal of the United States,* 19(2):1–5.

CHAPTER 2

THE SOUTHERN OCEAN

The shining surface of the water lies flat and mirror-like under the angled rays of a morning sun. There is a gentle puff of breeze. The air slides smoothly over the flat pan of water as if it were oiled. The water does not move. Now the breeze intensifies a bit and, at a point, the movement of air closest to the water becomes turbulent and ricochets off the surface, imparting some of its kinetic energy to the flat surface. The water ripples, and a "cats paw" moves over its surface, trailing the puff of wind. As the breeze picks up, more and more of its energy is pressed into the water, which now is highly irregular as ripples grow to wavelets that become waves. The waves roll before the wind with a speed and height that are influenced by conditions that include the depth of the water, the intensity of the wind, the distance which the wind is in unbroken contact with the water (fetch), the temperature differential between the water and air, the viscosity of the water, and a host of others. The physics of wind-driven waves, their formation, and the water currents which they generate are complex. But it is an important complexity, for the wind and its interactions with the Southern Ocean are critical for the vast majority of organisms in the Antarctic. If the winds stopped, so, eventually, would much of the life in the Southern Ocean.

Ten percent of the world's ocean consists of the Southern Ocean—the vast circumpolar windy and rough sea that entirely surrounds and isolates Antarctica. More than any other single body of water, the Southern Ocean dominates meteorological and biological processes occurring throughout its adjacent continent. Yet, unlike the austere, cold, dry land, this ocean is an extraordinarily rich and biologically productive place, with ecological relationships unlike any other. To begin to understand this richness—and to appreciate the many mysteries that still lurk here—we should delve into its physical oceanography; the system of winds, waves, currents, water chemistry, and tides that is the basis of all its life.

ANTARCTIC CONVERGENCE

The Southern Ocean has an indistinct northern boundary that at the surface is marked by the Antarctic Convergence. This meandering rim of the ocean is a place of dynamic interaction where cold antarctic surface water, flowing north, meets warmer subantarctic surface water flowing south from the Atlantic, Indian, and Pacific Oceans. In this zone of convergence the warmer, more highly saline northern water mixes with the colder, less salty antarctic water. The resulting mixture is denser than either of the parent water masses, and sinks at the convergence to form the Antarctic Intermediate water that slides slowly northward, about 800 meters below the surface layer.

The large-scale dynamics of sea water are profoundly affected by its density or specific gravity. Cold sea water is denser than warmer water; and the more salt it contains at a given temperature, the denser will it be. Fresh water is densest at about 4°C, and becomes less so at progressively lower temperatures. Sea water does not behave this way—it becomes increasingly dense with lowering temperature. Thus, the mixed water at the Antarctic Convergence is heavier than either of its parents and sinks to an intermediate depth.

The latitudinal position of the Antarctic Convergence varies according to longitude, season, and weather conditions. It can loop, swirl, and eddy; and in its peregrinations vary as much as 150 kilometers north and south. Despite this variability, the Antarctic Convergence generally is found in the neighborhood of 50° South latitude. Here, its presence is signaled by a dramatic drop in sea surface temperature as one sails south; from 8°C above the convergence to less than 2°C south of it. To bygone sailors of whaler, sealer, and clipper ships, that drop in sea surface temperature (and a corresponding one in the air above it) meant they were entering the windiest and roughest ocean in the world—a fact just as true today as then.

Below the Convergence the area of the Southern Ocean encompasses

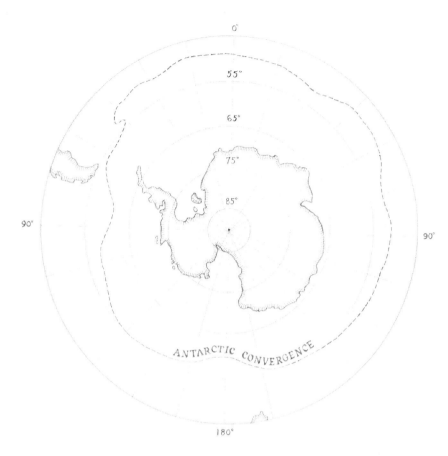

about 36 million square kilometers. Its dynamics are profoundly affected by a number of factors. These include wind direction and intensity, air temperature, bottom and shore topography, and the annual cycle of sea ice production and dissipation. It is the production of sea ice, particularly during the austral fall and winter, that gives the Southern Ocean a character of its own, and an influence on all other major oceans of the world.

ICE

The minimum sea ice cover of the Southern Ocean occurs in March, when only 4 million square kilometers of ice exists (a little more than 11 percent of the ocean's total area). With the onset of winter weather, the production of new sea ice, both "pack" ice (shifting floes) and "fast" ice (solidly fused to shore), produces a maximum ice cover by September of about 20 million kilometers of ice (57 percent of the ocean's total surface area). If this 16 million square kilometers of new ice averages 1.5 meters in thickness, the annual production of ice approximates 12 thousand cubic kilometers. During the process of forming this vast amount of ice there are salt concentration changes that are especially consequential in the Antarctic.

Pure sea water of salinity 35 parts of salt per thousand parts of water (35°/oo) freezes at a temperature of about −1.9°C (28.8°F). When sea water freezes, it is actually only the water fraction that crystallizes. The salts are left behind, there being no room for them in the growing lattice of water molecules. Platelets of ice (2.5 centimeters by 0.5 millimeters) grow that are initially surrounded by a brine of liquid water enriched by the excluded salts. This "frazil" ice floats to the surface, or packs against the underside of existing ice cover. If further cooling occurs, more water is frozen and a brine of increased concentration (and reduced volume) results. Frozen platelets of water ultimately become connected to each other to make larger crystals of ice. If this final freezing happens very slowly, the pockets of brine between adjacent platelets are gradually squeezed out and the dense, salty solution drains beneath the growing ice. The final salt concentration (salinity) of such ice is low.

On the other hand, when sea ice is rapidly formed, substantial pockets of brine are trapped within the ice, and their sequestration results in ice with different mechanical properties and melting characteristics. Rapidly frozen sea ice can have a salinity as high as 20°/oo. Most sea ice, however, has salinities that range from 10°/oo down to 2°/oo, with the average lying closer to the lower value.

To the pioneering arctic and antarctic explorers an understanding of the types of sea ice and their characteristics was of great necessity when sailing, sledging, or walking in the pack. The first stages of freezing, when platelets of frazil float to the surface, produces "slush" or "sludge" ice. If a newly formed sheet of ice is broken into fragments that become small circular floes with slightly raised rims from bumping against one another, "pancake" ice results. "Young" ice is unhummocked and has a fibrous consistency due to the unconsolidated frazil in it. Up to a foot thick, it breaks easily and presents little resistance to a ship sailing through it. When young ice results from the freezing of a smooth sheet of water under calm conditions, it sometimes is known as "black" ice.

Fast ice is held to some feature of the coast. It often carries a heavy snow burden and is unhummocked. An ice "field" is an expanse that is unbroken as far as the eye can see; while a "floe" is a free floating area of ice whose complete extent is visible. "Hummocky" ice is that which has been thrown into ridges by internal expansion or by forces generated by water currents and winds. It is usually old ice from which much of its brine has drained off, and has a translucent appearance with a spotted or "bubbly" look to it.

"Drift" ice occurs when a solid pack of ice breaks up so that the area of water exceeds that of the ice. Ocean swells occur here and contribute

"We were in pack ice from this past winter—one year's worth. Ice floes a meter thick and pancake ice in between. It forms in circular pieces that bump into each other, raising the edges where they hit."

"The sea is mostly covered with white pack ice speckled with very blue chunks of 'old ice.' Patches of open ocean appear a very dark steel grey as the ship plows through."

to the further degradation of the ice cover. The final product of this de-terioration is "brash" ice—the small bits of icy detritus that signal the final demise of the drift ice.

A large amount of ice in the Southern Ocean originates in the glaciers of Antarctica. Calving from glacial streams or cracking from the large ice shelves, this adds about 1.3 million cubic kilometers of ice to the ocean each year. The bergs are gradually transported northward as they sweep

"Much ice has been calving off lately for some reason—I've heard four icefalls in the past two hours. They look small next to the huge glacier face, but they create some big chunks. There's one here the size of a house."

around the continent in ocean currents. Most of them survive for from four to six years before melting.

Icebergs erode from the action of water waves and currents that often sculpts fantastic shapes in them. The blue and green colors of solid glacial ice are continually altered by the ever-changing light conditions of the Antarctic, which produce the kaleidoscope of subtle pastels and vivid colors that make icebergs marvels of beauty. Because bergs often have dangerously projecting underwater shelves, and especially because they frequently roll over as their centers of gravity shift due to subsurface melting, this beauty is best admired from some distance.

As icebergs melt they reduce to "bergy bits"—house-sized chunks—over which the sea breaks, completely washing the snow from them. They finally end their existences as "growlers"—pieces of green ice, barely awash at the surface, that present a particular menace to smaller ships navigating in the Southern Ocean.

WATER MASSES

The production of 24,000 cubic kilometers of ice in the Southern Ocean has a tremendous salt-enriching effect on the water under the forming sea ice. The highly saline, cold water produced under the winter ice pack

"The iceberg is full of patterns—honeycomb grooves, and layers and ridges—showing its original orientation when it was formed."

sinks, slides over the continental shelf and down into the abyss. As it goes it mixes with a tongue of a deeper water mass that projects south all the way from the North Atlantic Ocean and, to a lesser extent, from the Pacific and Indian Ocean basins. The dense surface water and deep water it mixes with form the coldest, most dense mass of sea water in the world— Antarctic Bottom Water. At temperatures close to $-1.8°C$ and salinities greater than $34.51°/oo$, the Antarctic Bottom Water sinks below the warmer Circumpolar Deep Water, moves to the west around the base of the continent, and slowly projects to the north where it can be identified in all of the major ocean basins of the world, including the North Atlantic well above the equator.

The Weddell Sea, particularly on its western side adjacent to the Peninsula, is where most of this Antarctic Bottom Water is formed. Here, in a vast area of winter sea ice formation, the bottom topography allows the dense, cold water to drop down and off the shelf, mixing with the warmer Circumpolar Deep Water. In other places of extensive ice cover, such as the Ross sea, troughs in the continental shelf trap the descended bottom water, keeping much of it from toppling off the shelf into the abyss.

As spring turns to summer over the Southern Ocean, increasing sunlight causes the extensive cover of ice to melt. Sea ice two or three meters thick can completely melt inside of six weeks. The immediate effect of the melt of low salinity ice is to freshen the surface water, which then floats on top of the more saline (but warmer) Circumpolar Deep Water. The Antarctic Surface water mass is characterized by its relatively low salinity (enhanced by precipitation, the melt of freshwater icebergs originating from the antarctic ice sheet, and sea ice melt) and its low temperature. This layer of water is shallow by oceanographic standards, being only 150 to 250 meters thick. Against a usual depth in the Southern Ocean of from 3,660 to 5,500 meters this surface water layer is paper thin. The bottom half of it is always close to the freezing point. In the north, near the Antarctic Convergence, the upper half may warm two or three degrees above zero Celsius.

The remaining major water mass found in the Southern Ocean is the

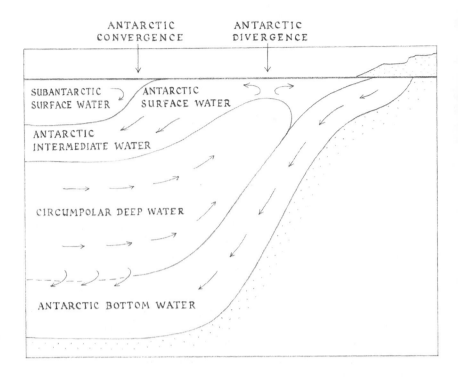

ANTARCTIC
CONVERGENCE

ANTARCTIC
DIVERGENCE

SUBANTARCTIC
SURFACE WATER

ANTARCTIC
SURFACE WATER

ANTARCTIC
INTERMEDIATE WATER

CIRCUMPOLAR DEEP WATER

ANTARCTIC BOTTOM WATER

Major water masses of the Southern
Ocean

so-called Circumpolar Deep Water (also known as the Warm Deep Water).
The latter designation is because the temperature of this thick, robust mass
of water is as high as +2.5°C just south of latitude 50° South, under the
Antarctic Convergence. The bulk of this water mass is formed from highly
saline water that sinks in the far North Atlantic. As it moves slowly south
it mixes with warmer, high salinity water that spills out of places like the
Mediterranean Sea deep below the surface currents at Gibralter. This Warm
Deep Water (found 1,500 to 3,000 meters below the surface) slowly moves
to the high southern latitudes where it is pushed upward by the Antarctic
Bottom Water moving northward beneath it. Although some Warm Deep
Water is received from the Indian and Pacific basins, most appears to
originate from the Atlantic. Altogether this layer represents the only source
of water to replace that lost from the Southern Ocean in the Antarctic
Deep and Antarctic Surface layers as they flow northward.

The temporal pattern of large-scale processes in the deep ocean is on a
different beat from that measured by human lives. And this, it turns out,
has consequences for our understanding (or lack of it) of biological processes in the Antarctic. A key tracer in timing the rate and past history of
biological and chemical processes involves measurement of the absolute
amounts of a radioactive isotope of carbon, ^{14}C, found in biological tissues.

The ocean gets most of its ^{14}C in carbon dioxide from the atmosphere, and the ratio of $^{14}C:^{12}C$ in a water mass can suggest how long it has been sequestered from direct contact with air. ^{14}C measurements tell us that the Circumpolar Deep Water is very old indeed, with apparent ages many hundreds of years since its last "deep breath" at the surface. Organisms that incorporate this "old" CO_2, or eat plants or animals that have, will appear to be extraordinarily old too. Tissues taken from a living penguin may seem by ^{14}C dating techniques to be hundreds of years old. Because of the ocean's relative lethargy, an important dating tool is lost to antarctic researchers.

WINDS

Superimposed on the larger system of water masses in the Southern Ocean is a complex array of currents and upwellings. These are mainly driven by winds and are responsive to the topography of Antarctica and its continental shelf. Preeminent among the features that contribute to the prevailing current patterns is the Antarctic Peninsula.

Because of the vast circumpolar fetch around the nearly circular continental outline, the winds of the Southern Ocean are uncomplicated. They exist in two concentric bands separated by a low pressure trough that cuts through the Peninsula in the vicinity of 65°S latitude. To the north of this the winds are usually from the west. The winds in the southern zone blow mainly from the east, although their proximity to the continent with its katabatic and convection winds makes this zone more variable. The mean wind velocity in the zone of the west winds is between 6 and 7 meters per second (13 to 15 miles per hour). Intense low pressure systems frequently spawn gale force winds and those of hurricane force can be expected several times a year, particularly in winter. The Southern Ocean is thus the windiest major body of water in the world.

WATER CURRENTS

The power of the winds provides the impetus for a system of currents in the Southern Ocean that has impressive biological effects. In the zone of westerly winds, water is driven mainly to the east with a pronounced northerly component (a current vector pushes surface water to the left of the wind direction in the Southern Hemisphere). The current is circumpolar and is known as the West Wind Drift. It extends through the Warm Deep Water mass to the bottom. The volume of water set in motion by the wind is immense, as much as 139 million cubic meters of water per second. In the open sea this current flows at about 8 miles per day (15 centimeters per second). In the Drake Passage the West Wind Drift is squeezed into the 800 kilometer stretch between the Antarctic Peninsula

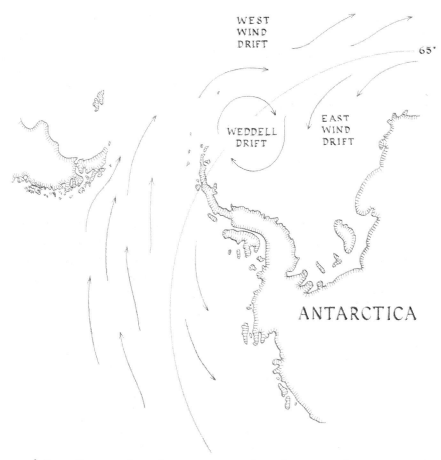

Surface currents in the Southern Ocean

and Cape Horn, and carries more water than does any other current system in the world.

South of latitude 65°S the easterly winds push the Antarctic Surface water to the west and south, causing a weaker and more complex current that narrowly circles the continent as the East Wind Drift. A consequence of the oppositely blowing major wind systems is to move surface water away from the narrow zone between them. Thus, surface water is pushed to the north in the West Wind Drift and to the south in the East Wind Drift. The gap between (known as the Antarctic Divergence) is filled by water from the Warm Deep mass which wells toward the surface. Although the Warm Deep Water does not actually reach the surface, it rises to within 100 meters of it in the Divergence; and this is shallow enough to put its nutrients within reach of phytoplankton in surface layer. The Divergence, thus, is a region of significant upwelling that stimulates the biological productivity of the Southern Ocean.

Snuggled against the Antarctic continent, the East Wind Drift is influenced by the topography of the continental shelf, the shoreline, and by the dynamics of icebergs released from the antarctic ice sheet. The most significant of these features is the Antarctic Peninsula. Projecting northward beyond the 62nd parallel, the Peninsula almost blocks the East Wind Drift and deflects most of this current northward where it reflects to the east into the bosom of the West Wind Drift. The gyre so produced is known as the Weddell Drift. Other lesser gyres spin off the East Wind Drift where it caroms off projecting features like the eastern margin of the Ross Sea, and smaller ice shelves that project from East Antarctica.

All of these gyres, but particularly the Weddell Drift, bring nutrients gleaned from the continental shelf into the more ice-free regions of the Southern Ocean. Consequently, some of the most biologically productive parts of this sea are coincident with them.

WAVES

The current structure of the Southern Ocean is not the only feature impressed upon it by winds. Most dramatic to our senses are waves that are an everlasting feature of this windy, storm-plastered ocean. Words like, ''windy,'' ''rough,'' ''wave,'' ''swell,'' ''comber,'' do little to convey the awesome power of the Southern Ocean. In 1916, Sir Ernest Shackleton undertook one of the most dramatic boat trips in history. He and his crew of the *Endurance* were stranded on Elephant Island (off the northern tip of the Peninsula) after their ship had been crushed and lost in the Weddell Sea ice pack. In order to save his crew (which miraculously he did) Shackleton and five of his men sailed more than 1,200 kilometers across the Southern Ocean in a 20-foot whale boat, protected only by a jury-rigged canvas cover. Shackleton used these words to describe the state of that sea:

. . . there were days and nights when we lay hove to, drifting across the storm-whitened seas and watching . . . the uprearing masses of water, flung to and fro by Nature in the pride of her strength. Deep seemed the valleys when we lay between the reeling seas. High were the hills when we perched momentarily on the tops of giant combers. Nearly always there were gales. So small was our boat and so great were the seas that often our sail flapped idly in the calm between the crests of two waves. Then we would climb the next slope and catch the full fury of the gale where the wool-like whiteness of the breaking water surged around us.[1] (*South!* [New York: Macmillan, 1920 and 1962], p. 168)

The enormous waves of the Southern Ocean are consequences of several factors—some of them unique to that sea. First, of course, is the force of the wind, blowing at a mean velocity matched nowhere in the world, and frequently accelerating to gale proportions or higher. The longer the wind is in contact with the water, the greater is the amount of energy imparted—and the longer the wave. In the Southern Ocean, particularly

in the West Wind Drift, that contact is virtually an unbroken one as the winds race clockwise around the continent. The fetch is essentially infinite. The transmission of energy from air to water is enhanced by the temperature differential between them. In situations where the air is colder than the water (the usual case in the Antarctic) the air is even more unstable and this facilitates energy transfer. Some studies suggest that wave heights may increase as much as 10 percent per degree Celsius of temperature differential. The same wind velocity, blowing in tropical and polar latitudes, generates quite different waves.

The energy content of waves is large. It consists of the kinetic energy of the motion of water within the wave, and the potential energy of the water raised in the wave to a height above mean sea level. All of this energy is spent when a wave exhausts itself in the surf zone. The energy content of an ocean wave two meters high (diminutive by antarctic standards) is equivalent to about 4,000 kilowatts per 100 meter length of wave. It is little wonder then that waves are such deterrents to shipping and to human development in the coastal zone. The energy content of the waves of the Southern Ocean dwarf the entire power production of our industrial society!

Waves generated by winds move at velocities that are proportional to their wavelength. Very long waves, those on the order of 350 meters and periods of about 15 seconds, are known as groundswells. In deep water, under calm conditions, their height may not be great, but they travel long distances at velocities approaching 25 meters per second. Groundswells generated in the Southern Ocean ripple northward through the South Atlantic, Indian, and Pacific Oceans, traveling literally thousands of kilometers. These groundswells build in shoal waters, concentrating their enormous energies, which then are expressed on the shorelines of exposed coasts throughout the Southern Hemisphere.

Shorter ocean swells and waves generated by westerlies in the Antarctic have periods of from 7 to 10 seconds and velocities of between 10 and 16 meters per second. Waves like these and the larger groundswells have great local consequences for mixing processes in the sea and thus the distribution of nutrients and organisms in the water column. In general, the particle motion in a wave is circular, with a radius of half the wave height at the surface. As one descends beneath the wave, the radius of the orbital movement decreases by about one half for each one ninth of wavelength in depth. For an ocean wave having a wavelength of 150 meters and a surface height of 8 meters (not untypical of Southern Ocean swells), a particle of water 75 meters deep would move in a radius of more than a third of a meter with the passing of each wave. This motion considerably aids the transport of deep nutrients to the sunlit surface area. No other ocean has such a well-mixed surface layer.

NUTRIENTS

The basis of nearly all life in the sea—and on land for that matter—is the production of complex organic molecules by green plants that use sunlight as an energy source to fuel the process. In addition to the water, carbon dioxide, and sunlight included in the basic overall photosynthetic reaction, plants need other chemical elements to flesh out the carbon-hydrogen-oxygen skeletons of their organic molecules. Nitrogen and sulfur are needed to build amino acids and their important polymers, the proteins. Phosphate and nitrogen are needed for nucleic acids. Potassium, calcium, silica, and several other inorganic constituents are vitally necessary for the construction of an array of other molecules that are necessary for life. If even one of these "nutrients" is in short supply, the production of new living matter will be curtailed. All the sunlight in the world cannot result in high biological production under such circumstances.

In terms of the major nutrient salts—phosphate, nitrate, potassium, silicate (the last needed by diatoms to construct their frustules (chapter 3)—the Southern Ocean is one of the richest in the world. South of the Antarctic Convergence the surface waters everywhere contain levels of these chemicals that equal or exceed the maximal concentrations (that usually occur only in the winter season) found in other oceans.

The nutritional requirements of the phytoplankton that convert sunlight to chemical energy range beyond the major phosphate, nitrate, and potassium salts. A host of trace elements are also needed for good growth, although not much is known about the exact requirements for most species found in the Southern Ocean. Marine algae in general also require growth factors such as vitamin B_{12}. These vitamins are generated by bacterial metabolism in the water column and sediment. And like nutrient salts, these are transported to the euphotic zone through the upwellings, current gyres, and turbulence that are hallmarks of the Southern Ocean.

TIDES

Ocean tides are essentially long period waves generated by and responding to gravitational effects of the earth, the moon, the sun, and, to a lesser degree, the distant planets. The actual amplitude of the tidal wave measured at any point is a complex function of the foregoing factors and the geometry of the basin where the tide is measured. Antarctic tides are difficult to measure, since the shallow regions in which they might be most apparent are so often ice-covered. Tidal amplitudes under ice cover such as the Ross Ice Shelf vary seasonally and daily as do tides through the world. But the normal maximum tidal amplitudes are less than one meter, with minimal ranges of only a few centimeters.

Local conditions, such as where partial blockage of tidal passageways by ice occurs, produce strong tidal currents. Nowhere is this more strongly expressed than in some of the recorded exploits of many of the early antarctic explorers. Open and inviting leads in the pack that suddenly (and sometimes catastrophically) slam shut; ice-bound boats that are carried to and fro by tidal currents; tidal bores that rush through openings in pack ice—all these and more attest to the importance of tidal currents on the fringes of the Southern Ocean. Tides are examples of yet another turbulent factor in the mixing of this fascinating sea.

CONTINENTAL SHELF

Except in the extensive Ross Sea and Weddell Sea areas, the continental shelf of Antarctica is narrow. The slope that drops rapidly to the abyss is within 100 to 200 kilometers of the coast nearly everywhere. As noted in chapter 1, the great weight of the antarctic ice sheet has depressed the entire continent so that the depth of the shelf is close to 400 to 500 meters. This is in striking contrast to other continents not burdened by ice, where the maximum depth of water over the continental shelf averages about 200 meters.

Sedimentation of the antarctic shelf appears to be low. This is mainly because of the absence of liquid water runoff from the continent, which would otherwise contribute silts and fine sands to the shelf. Instead the bottoms of the antarctic shelf typically display sediments in the form of rocks and coarse gravel that represent glacially transported material from the continent. The apparent fact that the ice sheet has been intact since the Miocene (perhaps 20 million years ago) means that reduced sedimentation has long been a feature of the shelf area. Deposition of sand and

gravel by glaciers at some distance from the present shorelines suggest that the ice sheet was, in the past, considerably extended from its current position.

A second factor that has conditioned the bottom sediments is the presence in many areas of currents associated with tides and the East Wind Drift that effectively pick up and transport finer sediments. This is true in the straits which separate the Peninsula from the South Shetlands and other islands of the Scotia Arc. Finally, at depths down to 15 meters, ice scouring is an important feature in recycling bottom sediments. Along unprotected shorelines ice scouring removes virtually all organisms and fine sediments down to five meters or so. Below this, down to 30 meters, there is a zone where super-cooled water freezes as it contacts the bottom. This "anchor" ice forms plates 20 or 25 centimeters by .5 centimeter and builds to half a meter or more in thickness. As its salts are expressed, the less dense ice rises to the surface along with its burden of entrapped sediments (and organisms). The anchor ice is then carried offshore in pack ice where, upon melting, its load is dropped into deeper offshore water. The absence of riverine sediments generally means that the antarctic benthos, particularly at depths below the zones of ice scouring and anchor ice formation, are "clean." Sessile organisms living there are not threatened by being buried in a choking hail of fine sediment. At depths below 30 or 40 meters, therefore, benthic organisms form rich mat-like communities (chapter 5).

While much of the shoreline of East Antarctica is steep, fairly regular, and features a ring of fast ice clinging to it, the shoreline of the Antarctic Peninsula is quite different. The complex geological history of this region

includes tectonic movement of as many as three microplates with attendant volcanic activity, plus substantial erosion. This created considerable relief in the topography of the shoreline and the submerged regions. The shoreline is jagged. Islands, large and small, are numerous. On the eastern, Weddell Sea side of the Peninsula, ice shelves and pack ice are predominant factors in scouring the bottom and inhibiting direct observation of it. On the western side of the Peninsula, in the lee of the prevailing easterly winds, ice is less predictable, and strong currents rushing through the island-strewn straits keep bottom sedimentation to a minimum. Here, however, are protected waters, lying behind islets where pockets of silt and other fine sediments exist.

The seaward edge of the continental shelf is marked by the abrupt descent from the 500 meter-deep shelf to the abyssal plain which averages about 4,000 meters (2,000 fathoms) deep throughout the Southern Ocean. The predominant current is the cold Antarctic Bottom Water directed west and north, away from the continent. The usual sediments are diatomaceous oozes with included fragments from radiolarans, foraminifera, and other shelled protists. The monotony of the even bottom floor is broken only rarely by obvious topographic features. Areas of subduction of tectonic plates produce the ocean trenches that are the deepest depths in the world ocean. In the Southern Ocean the deepest spot lies in the South Sandwich Trench, 9,038 meters—about 27,114 feet deep. The relief of this southern end of the world thus ranges from 9,000 meters below sea level to 5,100 meters above in the Vinson Massif—a total of 14,100 meters, or about 42,300 feet!

From a functional standpoint the most important topographic feature of the offshore region of the Southern Ocean is the Scotia Arc, a ridge which sets across the West Wind Drift at a depth of about 2,000 meters. The Scotia Arc passes eastward from southern Argentina before bending south and then west to terminate at the northern tip of the Antarctic Peninsula. At several places islands project from this ridge. Prominent ones include: South Georgia, Deception, Elephant, the South Sandwich, the South Orkneys, and the South Shetlands. Recent volcanic activity is known from some of these islands (Deception Island, the South Shetlands), but the Scotia Arc is not a ridge of the active sea floor spreading type.

The Arc, with its ridges and islands, forms a funnel—the Drake Passage—with a diameter of less than 800 kilometers through which all of the water of the West Wind Drift is forced. The current velocities are highest here, and turbulence is abetted by the islands and the irregular bottom topography. The mixing and vertical transport of nutrients here results in high levels of biological productivity.

The Southern Ocean is like no other. Its winds, waves, and currents are unique. Its levels of nutrients are matched by no other ocean. Its dynamics are pulsed in phase with the predominant feature of reduced solar radiation in the winter. This fact produces the extensive winter-spring ice cover that, in turn, is responsible for the production of cold, saline bottom water in the winter; and, only slightly warmer, less saline surface water in the summer. Given the existence of zoned bands of winds that blow across an infinite fetch, the ingredients are here for conditions that can favor enormous blooms of phytoplankton when the sun does shine.

It is that process of photosynthesis, both on land and in water, that drives the living world. All natural history stems from it. That process in the Antarctic will make a suitable subject for the next chapter.

ADDITIONAL READING

Anderson, J. B., C. Brake, E. W. Domack, N. Myers, and J. Singer. 1983. Sedimentary dynamics of the Antarctic continental shelf. In R. L. Oliver, R. R. James and J. B. Jago, eds., *Antarctic Earth Science,* pp. 387–389. Cambridge: Cambridge University Press.

Barber, N. F. and M. J. Tucker. 1962. Wind Waves. In M. N. Hill, ed., *The Sea,* 1:664–699. New York: Interscience John Wiley.

Dayton, P. K., G. A. Robillard, and R. T. Paine. 1970. Benthic faunal zonation as a result of anchor ice at McMurdo Sound, Antarctica. In M. W. Holdgate, ed., *Antarctic Ecology,* 1:244–258. London and New York: Academic Press.

Deacon, G. E. R. 1963. The Southern Ocean. In M. N. Hill, ed., *The Sea,* 2:281–296. New York: Interscience John Wiley.

Deacon, G. E. R. 1984. *The Antarctic Circumpolar Ocean.* Cambridge: Cambridge University Press.

El-Sayed, S. Z. 1970. On the productivity of the Southern Ocean. In M. W. Holdgate, ed., *Antarctic Ecology,* 1:119–135. London and New York: Academic Press.

Foster, T. D. 1984. The marine environment. In R. M. Laws, ed., *Antarctic Ecology,* 2:491–532. London and Orlando, Fl.: Academic Press.

Fu, L-L. and D. B. Chelton. 1984. Temporal variability of the Antarctic Circumpolar Current observed from satellite altimetry. *Science,* 226:343–346.

Gruzov, E. N. and A. F. Pushkin. 1970. Bottom communities of the upper sublittoral of Enderly Land and the South Shetland Islands. In M. W. Holdgate, ed., *Antarctic Ecology,* 1:235–238. London and New York: Academic Press.

Knox, G. A. 1970. Antarctic marine ecosystems. In M. W. Holdgate, ed., *Antarctic Ecology,* 1:69–96. London and New York: Academic Press.

Krebs, W. N. 1983. Ecology of neritic marine diatoms, Southern Harbor, Antarctica. *Micropaleontology,* 29(3):267–297.

Laws, R. M. 1984. *Antarctic Ecology.* 2 vols. London and Orlando, Fl.: Academic Press.

Laws, R. M. 1985. The ecology of the Southern Ocean. *American Scientist,* 73:26–40.

Robinson, E. S., R. J. Williams, H. A. C. Neuberg, C. S. Rohrer, and R. L. Ayers. 1975. Southern Ross sea tides. *Antarctic Journal of the United States,* 10:155–159.

Stein, M. and S. Rakusa-Suszczewski. 1984. Meso-scale structure of water masses and bottom topography as the basis for krill distribution in the S. E. Bransfield Strait, February–March 1981. *Meeresforsch.,* 30:73–81.

Trantor, D. J. 1982. Interlinking of physical and biological processes in the Antarctic Ocean. *Oceanography and Marine Biology Annual Reviews,* 20:11–35.

NOTE

1. Reproduced with permission from *South!,* by Sir Ernest Shackleton (New York: Macmillan, 1920 and 1962), p. 168.

CHAPTER 3

GREEN PLANTS—
THE PRIMARY
PRODUCERS

Very nearly all living processes, including the muscular work that it takes to write these words and the nervous energy consumed in reading them, are legacies of our star, the Sun. Sunlight is transformed in green plants through photosynthesis into the chemical energy of molecular bonds. This energy is then available for the metabolic reactions (respiration and growth) of plants, and of the animals which eat them. The terrestrial and oceanic ecosystems of the Antarctic depend on energy (primary productivity) converted from sunlight by both terrestrial and marine green plants. While the terrestrial component of this productivity in Antarctica is not large, the adaptations of plants there and their interactions with the oceanic sector are subtle and surprisingly complex. In the ocean, however, a rich array of photosynthetically important organisms exists on a scale quite different from the familiar plant life—cereal grains and fodder—that fuel life for most of us.

TERRESTRIAL PLANTS

In 1841 Joseph D. Hooker, an English botanist and early supporter of the young Charles Darwin, visited the Antarctic as surgeon and naturalist on

Sir James Clark Ross' *Erebus* and *Terror* expedition. He compiled the first list of terrestrial plants living there. In his catalogue of only 18 species Hooker reached conclusions that are as valid today as they were almost 150 years ago. That is, the terrestrial flora of Antarctica is dominated by what botanists refer to as cryptogamous plants—microscopic or small, often encrusting forms—that feature algae, lichens, fungi, and mosses. With the exception of but two species of flowering plants not seen by Hooker (a grass and a pink), the recorded flora today consists almost entirely of such cryptogams.

Terrestrial plant life on the Antarctic Peninsula has to deal with harsh conditions. Cold temperatures, low humidity, and lack of soil conspire to limit plant growth and reproduction. Especially important in this regard are the scarcity of liquid water and the absence of a soil-held reservoir of nutrients to support growth and development. The plants that do hang on to a precarious existence in this environment are those forms, exemplified by the mosses and lichens, that can cling to bare but protected rock faces which warm enough in the summer sun to produce trickles of melt water from snow patches or wind-blown ice crystals. They must also subsist on scarce nutrients leached from rock, or that are deposited on them by animals or by wind-blown dust, snow, and sea spray. The importance of this last factor is shown by the fact that the most luxuriant stands of lichens and mosses on the Peninsula are often found near the traditional rookeries of penguins, skuas, and other sea birds. Here nesting materials imported by the birds enrich the ''soil''; and powdered guano nourishes the nutrient-starved plants. Even the terrestrial plants are stimulated, in part, by nutrients (guano) delivered from the marine ecosystem.

The most equable and productive regions of the Antarctic Peninsula are the maritime western areas north of Latitude 70° South. The central, eastern, and southern sectors of the Peninsula are predominately snow and ice covered, and have environments more similar to that of East Antarctica. In the maritime regions the mean monthly temperature rises to above 0° Celsius at midsummer and the annual precipitation (in water equivalent) ranges from 25 to 100 centimeters, some of it as rain. A similar set of conditions exists in the South Shetlands, the South Orkneys, and other islands of the archipelago.

Under the best conditions the mass of living mosses and lichens grows richly with high production rates. At the height of the austral summer, on the infrequent clear days, the temperatures of the exposed rock faces on which these plants grow rise to as much as 40°C (although the air temperatures a few centimeters above may hover around 0°C). Under these conditions moss turves and carpets show good production. In moist areas the annual production often exceeds the rate at which consumption

by herbivores or decomposers occurs. The dead moss shoots and rhizomes accumulate, forming peat banks under the active turf. In some places the peat builds to a meter or more in depth, with all but the top 25 centimeters being permanently frozen.

During most of the rest of the year, however, these plants simply endure the cold, the drought, and nutrient deprivation, while resting in a

''The moss is a foot or so thick in some areas. Green spongy-soft on top with brown 'stems' running all the way down. Moist and dense in there; you can part it, there's nothing holding the clump together except the density of it. You can run your hand underneath a mat of it and it rises like a blanket.''

kind of suspended animation. Growth phases are transient and the ages of many cryptogams are measured not in years, but rather in centuries. Indeed, sexual reproduction in some antarctic mosses occurs so infrequently that it has seldom been observed by visiting scientists!

Lichens

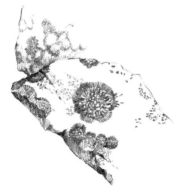

Crustose lichens

Lichens belong to a group of organisms that are familiar to us as common vestments of our country rocks and trees. Yet their structure, physiology, and life cycles are so complex that most biologists know little about them. A lichen is composed of a combination of two different life forms: one a photosynthetically active green (or sometimes blue-green) alga, the other a fungus—a non-photosynthetic consumer that in many classifications of life belongs to its own kingdom, quite distantly removed from the algae. This association grows and reproduces, and exists in recognizably discrete forms analogous to species. The working lichen is really a partnership in which the alga, through its photosynthesis, provides energy to itself and to the fungus; which returns the favor by providing the alga with the protection, moisture, and nutrients it requires. While the complexities of this relationship are still not completely understood by lichenologists, it can be seen as a successful and enduring experiment in evolution.

Lichens are among the first complex organisms to invade new and often harsh environments, paving the way for other sorts of plants and animals. In the Antarctic, however, conditions are so severe and the metabolic processes of the plants proceed so irregularly and slowly that the normal ecological succession has all but been arrested in its pioneer stage.

It is difficult to know just how many kinds of lichens exist in Antarctica, but about 100 species are thought to exist on the Peninsula. The crustose forms predominate, although fruticose and foliose lichens also are common. By far the richest stands of lichens live on the protected western side of the Antarctic Peninsula and its offshore islands where they provide splashes of gold and orange color in an otherwise stark landscape.

Fruticose lichen

Among plants, lichens respire and photosynthesize at the lowest temperatures, light levels, and moisture contents. They are thus well adapted to life in Antarctica. Most of the lichens found here grow exceedingly slowly, with diameters increasing at rates of from 10 millimeters to 16 millimeters *per century*. Estimates of net primary productivity range up to 250 grams (dry weight) per square meter per year.

Mosses

The bryophytes or mosses can survive under conditions nearly as extreme as can lichens, but they are less likely to adhere to bare rock—preferring

Fruticose lichen

the small patches of sandy soil found on the Peninsula and its adjacent islands. Like lichens, antarctic mosses thrive especially well near sea bird nesting colonies where soil emollients in the form of nesting materials (lichens, seaweeds, and flotsam) and guano enrich the ground.

In such places the net primary production in dense moss turves can average more than 400 grams of dry weight per square meter per year. This is higher than that measured in many northern and alpine tundras, and underlines the capacity of the Antarctic to be rich. Unfortunately such terrestrial plant production is confined to a tiny percentage of the total land area, and its ecological contribution is therefore small.

About 45 species of mosses are known from the Peninsula and its associated islands. Two typical and dominant species are *Polytrichum alpestre* and *Chorisodontium aciphyllum*. One question that intrigues botanists and zoologists alike concerns the origins of the species of plants and animals that occupy modern Antarctica. Are they derived from invaders from the the north—most likely South America—in the last ten thousand years; or are some of them relics of the ancestral antarctic flora and fauna which survived the extensive ice ages that have swept the continent since it arrived at its present polar position? Answers to this question must await more extensive and intensive floral and faunal analyses by specialists who compare associations of living species in different regions with those in the fossil record. Such scientists today are few in number, and their research is not richly funded, so the wait may be lengthy. If relic terrestrial species exist in Antarctica, however, they probably will be found among the mosses and lichens.

In addition to mosses, the Peninsula plays host to five species of a closely-related group of plants, the liverworts. Nowhere else in Antarctica have these small and unobtrusive plants been found. Like mosses, the liverworts practice a form of sexual reproduction where motile sperm-like cells move to the female reproductive organs that house the ova to be fertilized. This process requires a film of liquid water in which the male gametes can swim. The infrequency of rain and the irregularity of meltwater in the needed quantities means that reproduction in some species of bryophytes is not a regular affair. Fortunately, these are long-lived plants, and they are able to bide their time for tens or, perhaps even hundreds of years.

Algae

In addition to lichens, mosses, and liverworts there are numerous species of unicellular algae reported from Antarctica in general and the Peninsula in particular. Genera found here include cosmopolitan ones such as *Chlorella* and *Chlamydomonas* that are sometimes featured in introductory biology courses. The magnitude of their photosynthetic contributions to the

"The glacier face is getting more and more orange-brown in places. It's an algal bloom from spores that have been trapped in the ice for a long time."

terrestrial ecosystem is not certain, although it is probably slight. Of special interest are unicellular algae (cryoplankton) living and reproducing in snowbanks and glacier faces on the Antarctic Peninsula. Such patches of algae are confined in their distribution to coastal areas where the snow can become saturated with meltwater during the summer. As it melts, the snow that contains these algae appears to be stained red or green as the plants become exposed. Green algae, diatoms, and fungi have been recovered from such snow.

In local areas near penguin rookeries where uric acid is found, mats of macro algae such as *Nostoc commune* reach standing crops as high as 300 grams per square meter. *Nostoc* is a blue-green alga (cyanobacteria) and has the ability to fix nitrogen gas from air. Less dense aggregations of *Nostoc,* therefore, can be important associates of mosses and lichens to which they supply nitrogen in a usable form.

Flowering Plants

Although there are abundant fossil remains of a former rich terrestrial plant community on Antarctica, only two species of native flowering plants are today known here, and these only on the Peninsula south to about 68° 21' S. The most common form is a grass *(Deschampsia antarctica)* that grows in sheltered spots made habitable for it by the soil-forming moss-

lichen community. It can also pioneer in those few sheltered places where fine mineral soils exist. This grass is common enough so that patches of it are found in places. It grows most luxuriantly in the South Shetland and South Orkney Islands. *Deschampsia* extends northward on the South American continent (where Hooker collected it) to 38° South in the Andes. *Deschampsia* is thought by some botanists to be a relic of the old antarctic flora that was first eliminated from the continent by glaciation, but which has subsequently reinvaded it from the north.

The other antarctic flowering plant is a pink, *Colobanthus quietensis,* which is less common than the grass, but which occupies the same kind of habitat. A weed *(Poa annua)* was reported from Deception Island just to the north of the Peninsula in 1953 and survived there for several years. From time to time attempts have been made to introduce other species of plants in the Antarctic. These introductions have included such forms as kale and bluegrass. The history of introductions of exotic plants and animals throughout the world is that native, rare species are sometimes exterminated by the new invaders. The rigor of the antarctic climate has been too much for the species introduced here, however, and they fortunately have not survived. All in all, the energetic contribution that the native antarctic flowering plants make to the terrestrial antarctic ecosystem must be minuscule at best; less, probably, than any other photosynthetic element in the flora.

Deschampsia antarctica

FRESHWATER LAKES

Algae and other unicellular organisms are found in meltwater pools that occur briefly in the summer, as well as in the water column and bottom

X 20

X 1

"COLOBANTHUS, the only other flowering plant besides the grass, DESCHAMPSIA. It grows very low to the ground, with small leaves."

communities (benthos) of freshwater lakes on the Peninsula. Most of the lakes that exist here and on the islands of the Scotia Arc have not been studied. Most of the ones that are well-known exist on Signy Island in the South Orkney Islands. These range from clear, nutrient-poor, oligotrophic lakes to eutrophic ponds that are enriched by sea spray, bird, and seal activity.

The plant life in these lakes consists mainly of green algae, mosses, and blue-green algae. The more oligotrophic lakes have sparse amounts of phytoplankton in the water column, but possess surprisingly rich mats of benthic mosses, green algae, and blue-green algae that coat the bottom from the maximum depth of winter ice (1 to 2 meters) down as deep as 7 or 8 meters. The winter respiration of these mats and that of other organisms in the benthos cause severe oxygen depletion of the water, and the deepest layers routinely become anoxic late in the winter.

Such mats or "felts" of benthic mosses and algae are less well-developed in the more nutrient rich mesotrophic and eutrophic lakes. Here phytoplankton populations are denser and the water is less transparent to light. In the summer, when the surface ice has melted, these lakes are

well mixed by wind-driven currents. Under ice cover, however, calm conditions prevail and the plankton settle out of the water column to the bottom.

Some lakes in the central and eastern parts of the Peninsula never lose their ice cover during the summer, and others exist under glaciers, being fed by basal melt water. The permanently frozen lakes sometimes support active photosynthetic and consumer communities, especially if winds keep the ice surface free of snow and allow sunlight to penetrate to the underlying water. Because these permanently frozen freshwater lakes are essentially sealed microcosms, they present unique opportunities for ecologists to study contained, reasonably simple ecosystems. That study has only just begun.

PRIMARY PRODUCERS OF THE SEA

Life in Antarctica is largely driven by photosynthetic events going on in the Southern Ocean. Here organisms are protected from extreme climatic and physical conditions by coverings of ice and water. Here also plants are nourished by abundant nutrients, so they can convert sunlight to chemical energy pretty much unimpeded by the limits often imposed on phytoplankton in temperate and tropical seas. The oceans of the world play host to a varied assemblage of minute, drifting plants (the phyto-

"I found a thin crack in the glacier. You can hear running water underneath and see it through the crack. A foot or two down there's a little under-ice stream of glacial meltwater that feeds our lake."

plankton) in which groups known as dinoflagellates and diatoms often predominate. It is one characteristic of the Southern Ocean that diatoms by far and away outrank the dinoflagellates in terms of biomass and photosynthetic importance. So vital are diatoms to the extraordinary animals that live in the Antarctic that it is worthwhile to discuss the biology of this group of organisms in some detail.

Diatoms

Diatoms belong to the kingdom of single-celled organisms known as the Protista. Within that kingdom they can be classified among the Phylum Chrysophyta—the golden algae. One of their outstanding characteristics is a cell wall strengthened by deposits of silica. The resulting shell or frustule is often fashioned into intricate designs typical of each species (there are about 10,000 species of diatoms—about 100 of which are found in the Antarctic) and modified by the environmental conditions under which it develops. Although each diatom is thus enclosed in a glass house, most of them are very tiny indeed, with diameters ranging from less than 10 micrometers up to a hundred or so (there are 1,000 micrometers in a millimeter). Most of the planktonic diatoms have diameters less than 50 micrometers, which means that they are too small to be captured with certainty in the finest meshed plankton nets. Their presences can be estimated by measurements of chlorophyll concentrations in water samples, or by centrifuging small volumes of sea water and then examining the confined sample directly under a microscope. The former technique fails to produce a species analysis, while the latter one is apt to overlook the rarer species present. Diatoms, important as they are, present real sampling and analytical problems to the ecologist.

Diatoms   Diatoms are found in all sorts of wet and sunlit environments. In addition to the drifting, planktonic ones, sedentary benthic forms adhere to

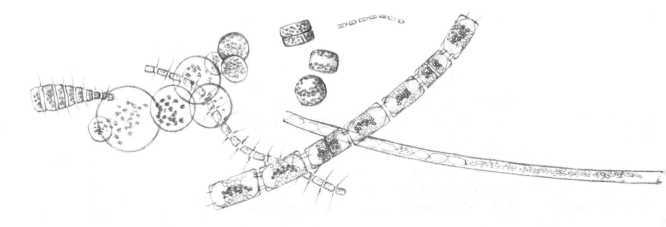

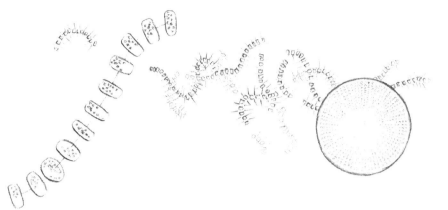

Diatoms

Size reduction in dividing
centric diatoms

Pennate diatoms

the bottom or to objects floating in the water above. The Blue Whale of antarctic waters (chapter 7) is also known as the Sulfur Bottom Whale because of the golden-brown color imparted to its belly by encrusting diatoms. Most importantly for the Antarctic, some diatoms live and grow while actually frozen into the pack ice that develops each winter.

Some planktonic diatoms are encased in frustules that are elongate (pennate diatoms) or are extended into eye-catching spines and processes. These probably give the cell more surface area to retard its sinking rate, thus keeping it in the sunlit zone where photosynthesis can take place. As another flotation device, some diatoms fail to separate completely following cell division, forming chains of daughter cells that present, together, a much larger surface area. Examined with a scanning electron microscope the frustules of diatoms are found to be elaborately perforated with tiny apertures that probably allow the direct contact of sea water with the underlying cell membrane, enhancing the uptake of nutrients from the water.

Another characteristic of the diatom frustule is its valvular nature. Nearly all diatoms are constructed of two overlapping half-shells, called valves. The smaller valve fits inside the larger much as one half of a pill box fits into the other. These are then fastened together by glassy girdle bands, which act as hoops holding a barrel together. This interesting anatomy has consequences for the average sizes of diatoms in succeeding generations, for when the cell divides, each new cell inherits one half of the old frustule. In both daughter cells the old half serves as the larger valve, with a new, smaller valve being constructed. Each generation is thus smaller than the generation that preceded it. With each cell division the average diameter of the population becomes smaller. This reduction in size continues until the diatoms of the population are only 10 to 50 percent of their maximum diameter. Then environmental factors such as changing temperature or light levels trigger the development of gametes and sexual reproduction takes place, restoring the new generation to its species-typical maximum diameter. This restoration often occurs in early spring, at the start of the new growing season. Under good laboratory culture conditions diatoms can divide as often as once per day. Under the most favorable conditions in the cold Southern Ocean the generation time is more like one division every three days.

Diatoms can be very numerous indeed, with some of the antarctic species found in concentrations as high as one million cells per liter of water. At these densities (blooms) the cells cause the otherwise clear sea water to be colored like mahogany. This hue is due to a brown pigment, fucoxanthin, that diatoms contain in addition to the more normal green photosynthetic pigments chlorophyll a and chlorophyll c. Fucoxanthin probably helps to trap light energy that during photosynthesis is then transferred to chlorophyll a to drive the actual incorporation of carbon dioxide into organic molecules.

Primary Productivity

Primary production (photosynthesis) requires not just sunlight, but also the nutrient salts needed to form enzymes and the structural molecules that help to make up living diatoms. Phosphates, nitrates, silica, potassium, and calcium are required by diatoms in fairly large quantities. The Southern Ocean, and particularly the region near the Antarctic Peninsula, presents these nutrients in abundant quantities. It is here that upwelling of the deep oceanic water at the Antarctic Divergence and the whirling gyres of the East Wind Drift bring nutrient-laden water to the sunlit surface zone. The considerable turbulence of the stormy Southern Ocean also helps in this cause.

One result of this enrichment is summertime algal growth and primary production as high as that found in any ocean in the world—this despite sea temperatures that hover around 0°C! Near the Peninsula and along the Scotia Arc primary production can reach 2 grams of carbon fixed per square meter of surface area per day. Most of the open Southern Ocean operates at a lower level, however, averaging perhaps 0.2 grams of carbon per square meter per day. Over a six-month growing season the primary productivity of the Southern Ocean is estimated to be about 650 million metric tons of carbon.

Because phytoplankton nutrients never seem to be scarce enough here to restrict either photosynthesis or cell division, early workers in the Antarctic thought that the Southern Ocean was the most productive on earth. It turns out that even though the phytoplankton here are well adapted to photosynthesizing in dim light, they do not completely compensate for the low temperatures of this sea. Temperature thus becomes a factor that depresses photosynthetic activity and keeps the overall average primary productivity on a par with most of the other oceans. Still, 650 million tons of carbon is roughly one tenth of the total net primary productivity of the watery world. The Southern Ocean is a very rich place indeed!

LIFE IN ICE—THE CRYOPELAGIC HABITAT

Diatoms and other forms of algae and microbes occupy a habitat in the Antarctic—ice—that is alien to our usual perceptions of living systems. In addition to his other original observations on plant life in the Antarctic, J. D. Hooker reported the presence of a rich assemblage of diatoms frozen into the sea ice fringing the continent. Recent studies demonstrate that these organisms don't just exist in the ice—they are able to photosynthesize and grow while entombed there. This special environment is called the cryopelagic habitat.

As sea ice forms, diatoms, other phytoplankters, and bacteria are entrapped in the frazil ice and are eventually frozen into the developing pack ice as the mass of crystals freezes solid in the fall. Many of the diatoms found frozen in this ice are typical of the species common to the open water. Measurements of chlorophyll a concentration in sea ice cores taken over time intervals show that these cells are actively photosynthesizing and growing during their stay in the ice pack, before it breaks up and becomes destroyed in the summer. Annual production rates estimated for this diatom community range up to about 50 tons of carbon per year.

In some areas of sea ice distinctive microbial communities exist that feature species clinging together to form long chains in tubes within the ice. Other species penetrate only the bottom centimeter or two of ice and hang in dense chains 5 to 10 centimeters into the water column. Many of the phytoplankters have abundant bacteria associated with them that could function as symbionts. In addition, a variety of heterotrophic protists and animals are also part of this community. These communities are quite fragile, and are so easily dislodged that careful study of them becomes difficult.

The phytoplankton assemblages sometimes become so dense that they stain the underside of the sea ice a rich brown. Amphipods and other animals accumulate here, presumably feeding on these cryoplankton, and they too may become entrapped in the forming ice. In the summer when the pack ice breaks up, Kelp Gulls and various petrels feed on the cryopelagic community as blocks of rotting ice overturn, exposing their rich flora and fauna.

Despite the intuitive picture most of us have about the undesirability of being frozen into ice, it turns out to be not a bad way for diatoms to spend some time. The ice, after all, is not colder than the underlying water with which it is in thermal equilibrium. The big advantage, of course, is that ice-bound algae are not subject to grazing by herbivores such as krill or copepods. As long as the diatoms can get adequate light and nutrients these cold-adapted organisms grow and divide. The sea ice community of diatoms is usually much denser than that of the water column beneath the ice. So successful are they that the concentrations of diatoms sometimes approach the theoretical maximum densities that they might be expected to reach in sea water! Of course these diatoms are restricted to just a meter or two of surface ice. The water column beneath is much thicker. Even though the grazed water community below is less populous, its total productivity dwarfs that of the ice community.

The ice community serves as a reservoir from which the surface waters are seeded as the pack ice melts. The grazed and dark depleted surface waters receive the nucleus of a new producer community each spring as the disintegrating pack ice spews its winter-entombed crop of diatoms into the open sea.

SEAWEEDS

In most places the littoral zone of the Antarctic is restricted by the near absence of a continental shelf. The Peninsula, however, has abundant shallow waters that serve as a substrate for seaweeds and microalgae. Green, red, and brown seaweeds exist in this environment, but do not seem to assume the kind of importance here that they have in more temperate waters. In shallow waters and tide pools mats of benthic diatoms develop during the summer and are grazed down by limpets that in turn support birds such as Kelp Gulls.

More than 75 kinds of seaweeds are known from the Peninsula and its associated islands. The richest flora is found on the exposed western coastlines where impressive stands of brown algae such as *Desmarestia anceps* exist. The more protected inshore water of the Peninsula, between islands and the mainland or in the lee of the prevailing winds, support only limited numbers of attached algae. This is because the turbidity of these protected waters, caused by suspended sediments and rich plankton blooms, restricts light penetration; and because of the presence of icebergs and winter pack ice. In these areas the bottoms down to about 10 meters in depth are heavily scoured by this ice, with some scouring evident as deep as 35 meters. This means that each summer's seaweed growth is scraped off by ice action.

Off the windward, more exposed coasts, wind and current action delay and reduce pack ice formation, and the clearer water allows benthic seaweeds to develop at greater depths. Consequently a more diverse, richer

Desmarestia

algal community is found there, some of it extending down to 100 meters or more.

Although the total contribution of seaweeds to the primary production of the Antarctic is but a small fraction of that of the pelagic and plank-tonic diatom community, these plants can provide energy for locally abundant communities of amphipods, isopods, fishes, and other organisms. They may be especially important in fueling detrital food chains dominated by bacteria.

This survey of the important primary producers of the Antarctic Peninsula and its nearby waters has dealt mainly with the lichens, mosses, diatoms, and seaweeds. The energy converted from sunlight to biomass by these organisms is the fuel that drives the rest of the living world of the Antarctic. We can now begin to explore the lives of those animals, learning how they use the energy made available to them; and how they in turn pass some of it along to other elements of this fascinating ecosystem.

ADDITIONAL READING

Balech, E., S. Z. El-Sayed, G. Hasle, M. Neushul, and J. S. Zaneveld. 1968. Primary productivity and benthic marine algae of the Antarctic and Subantarctic. *Antarctic Map Folio Series,* folio 10, pp. 1–12. New York: National Science Foundation, American Geographical Society.

Buck, K. R. and D. L. Garrison. 1983. Protists from the ice-edge region of the Weddell Sea. *Deep-Sea Research,* 30 (12A):1261–1277.

Davey, A. 1983. Effects of abiotic factors on nitrogen fixation by blue-green algae in Antarctica. *Polar Biology,* 2:95–100.

Davis, R. C. 1981. Structure and function of two antarctic terrestrial moss communities. *Ecological Monographs,* 51(2):125–143.

Dieckmann, G., W. Reichardt, and K. Zielinski. 1985. Growth and production of the seaweed, *Himantothallus grandifolius,* at King George Island. In W. R. Siegfried, P. R. Condy, and R. M. Laws, eds., *Antarctic Nutrient Cycles and Food Webs,* pp. 104–108. Berlin: Springer Verlag.

Ellis-Evans, J. C. 1984. Methane in maritime antarctic freshwater lakes. *Polar Biology,* 3:63–71.

Fenton, J. H. C. 1980. The rate of peat accumulation in antarctic moss banks. *Journal of Ecology,* 68:211–228.

Fiala, M. and L. Oriol. 1984. Vitamine B_{12} et phytoplancton dans l'Océan Antarctique. Distribution et approche expérimentale. *Marine Biology,* 79:325–332.

Garrison, D. L., C. W. Sullivan, and S. F. Ackley. 1986. Sea ice microbial communities in Antarctica. *Bioscience,* 36(4):243–250.

Grossi, S. M. and C. W. Sullivan. 1985. Sea ice microbial communities. V. The vertical zonation of diatoms in an antarctic fast ice community. *Journal of Phycology,* 21:401–409.

Hawes, I. 1985. Factors controlling phytoplankton populations in maritime antarctic lakes. In W. R. Siegfried, P. R. Condy, and R. M. Laws, eds., *Antarctic Nutrient Cycles and Food Webs,* pp. 245–252. Berlin; Springer Verlag.

Heywood, R. B. 1984. Antarctic inland waters. In R. M. Laws, ed., *Antarctic Ecology,* 1:279–334. London and Orlando, Fl.: Academic Press.

Heywood, R. B. and T. M. Whitaker. 1984. The marine flora. In R. M. Laws, ed., *Antarctic Ecology,* 1:373–419. London and Orlando, Fl.: Academic Press.

Holdgate, M. W. 1977. Terrestrial ecosystems in the Antarctic. *Philosophical Transactions of the Royal Society of London B.,* 279:5–25.

Holm-Hansen, O., S. Z. El-Sayed, G. A. Franceschini, and R. L. Cuhel. 1977. Primary production and the factors controlling phytoplankton growth in the Southern Ocean. In G. A. Llano, ed., *Adaptations Within Antarctic Ecosystems,* pp. 11–50. Washington, D.C.: Smithsonian Institution.

Ino, Y. 1983. Estimation of primary production in moss community on East Ongul Island, Antarctica. *Antarctic Record,* 80:30–38.

Jacques, G., M. Fiala, and L. Oriol. 1984. Démonstration, à partir de tests biologiques, de l'effet négligeable des éléments traces sur la croissance du phytoplancton antarctique. *Comptes Rendus de l'Académie des Sciences.* (Paris), Série III. 298(18):527–530.

Krebs, W. N. 1983. Ecology of neritic marine diatoms, Arthur Harbor, Antarctica. *Micropaleontology,* 29(3):267–297.

Palmisano, A. C. and J. B. SooHoo. 1985. Photosynthesis-irradiance relationships in sea ice microalgae from McMurdo Sound, Antarctica. *Journal of Phycology,* 21:341–346.

Sasaki, H. and K. Watanabe. 1984. Underwater observations of ice algae in Lüt-zow-Holm Bay, Antarctica. *Antarctic Record,* 81:18.

Smith, R. I. L. 1984. Terrestrial plant biology of the sub-Antarctic and Antarctic. In R. M. Laws, ed., *Antarctic Ecology,* 1:61–162. London and Orlando, Fl.: Academic Press.

Smith, R. I. L. 1985. Nutrient cycling in relation to biological productivity in Antarctic and Subantarctic terrestrial and freshwater ecosystems. In W. R. Siegfried, P. R. Condy, and R. M. Laws, eds., *Antarctic Nutrient Cycles and Food Webs,* pp. 138–155. Berlin: Springer Verlag.

Smith, W. O. and D. M. Nelson. 1986. Importance of ice edge phytoplankton production in the Southern Ocean. *Bioscience,* 36(4):251–257.

Tilzer, M. M., B. von Bodungen, and V. Smetacek. 1985. Light dependence of phytoplankton photosynthesis in the Antarctic Ocean: Implications for regulating productivity. In W. R. Siegfried, P. R. Condy, and R. M. Laws, eds., *Antarctic Nutrient Cycles and Food Webs,* pp. 60–69. Berlin: Springer Verlag.

Usher, M. B. 1983. Pattern in the simple moss-turf communities of the sub-antarctic and maritime antarctic. *Journal of Ecology,* 71:945–958.

Yamaguchi, Y., S. Kosaki, and Y. Aruga. 1985. Primary productivity in the Antarctic Ocean during the austral summer of 1983/84. *Transactions of Tokyo University of Fisheries,* 6:67–84.

CHAPTER 4

ANTARCTIC HERBIVORES

In the north, high-latitude ecosystems feature large numbers of relatively few species of plant-eating animals. This contrasts with the structure of ecosystems in temperate and tropical latitudes, where herbivores are usually more diverse both in number of species and in abundance. From the viewpoint of the theoretical ecologist this means that ecosystems in high latitudes are structurally more simple, and potentially easier to understand.

Where the Northern Hemisphere's major Arctic terrestrial ecosystem—the tundra—is extensive and relatively accessible, it also has a diverse herbivorous fauna associated with it that includes vertebrates such as birds, rodents, caribou (reindeer), and musk oxen. Much of this fauna has been altered by people in recent times. Indeed, humans are endemic to this habitat.

TERRESTRIAL HERBIVORES

Antarctic terrestrial ecosystems, by contrast, present a different picture. Aside from snow algae, most of the primary productivity is confined to lichen and moss communities that are luxuriant only on the northern part of the Peninsula and its islands. There are no large herbivores—nei-

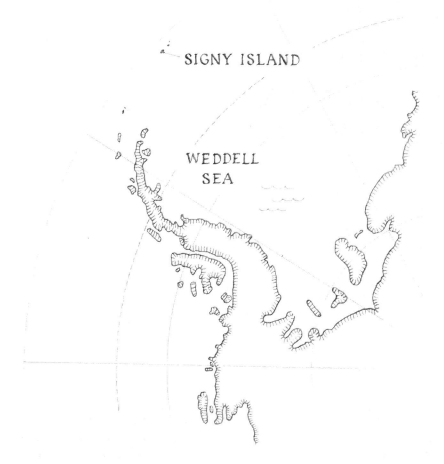

SIGNY ISLAND

WEDDELL SEA

ther bird nor mammal. Human intervention has been well controlled and, except for some general pollution of ocean and atmosphere, nondestructive.

Even though terrestrial ecosystems represent but an insignificant fraction of the overall biological productivity of the entire antarctic region, they are of great interest to ecologists who look to study simpler natural ecosystems. In this context the British Antarctic Survey has for several years supported an intensive study of the terrestrial communities on Signy Island in the South Orkneys that lie to the north and east of the northern tip of the Antarctic Peninsula. In concert with other projects carried out on the Peninsula and elsewhere on the continent, this study has provided a better understanding of the ecological interactions in this sparse environment.

The community associated both with mats of lichens and turves of mosses is not large. Here are found protists, rotifers, nematodes, tardigrades, and a few arthropods—some mites, springtails, and midges. Most of these groups include species that are carnivores or feed on detritus—so all major trophic levels are present.

Although simplicity is suggested by the few kinds of organisms, the study of this ecosystem is complicated by the small sizes of the animals in it. Many of the protists are but a few micrometers in diameter; rotifers average 0.25 millimeters in length; and adult tardigrades and mites may only reach 0.5 millimeters. The giants of the community are springtails (Collembola), the most common species of which grows to a maximum length of about two millimeters.

The factors that most severely limit the flow of energy through this ecosystem are the deadly duo of cold and dryness. Temperature, of course, limits metabolic activity; and most of the protists and animals in this system do not function much below the freezing point of water. All of them, however, can endure extreme cold due to supercooling, cryoprotectants, and other strategies. It is thus the presence of moisture as a buffer to desiccation that is an even sterner factor governing survival than is low temperature.

The isolation of Antarctica governs the kinds and distribution of herbivores and carnivores found in it. In the Arctic tundra, the land is contiguous with southern, more equable, climes. This allows easier invasion on one hand, and retreat from the brutal realities of winter on the other. Neither is possible for a land animal on the Antarctic Peninsula. Dispersal to this environment must be as aerial spores, by hitchhiking on migrating birds, or by riding flotsam. Survival through the winter can only come by hunkering down and bearing it. Overland emigration is not an option.

Protists and animals live anywhere in the Antarctic that liquid water can exist and a source of organic carbon can be found. Such places occur under the surface layers of talus slopes in the Transantarctic Mountains and isolated nunataks of East Antarctica; the fellfield gravel slopes of the Peninsula and islands; within lichen patches, moss turves, and among the roots of the isolated patches of *Deschampsia* and *Colobanthus*. Even under the most severe conditions of East Antarctica, mites and springtails exist in gravel "soils" at altitudes as high as 2,200 meters above sea level, where water concentrations in the gravel are as low as 2 percent. The richer and more interesting antarctic terrestrial ecosystem, however, is found in the moss turves and carpets of the Antarctic Peninsula.

In this habitat (using the Signy Island community as a model) one can find populations of protists that number as many as 900 million cells per square meter, with a total biomass in that area as high as 1.7 grams dry weight. In the wetter moss clumps rotifers are common. They feed on plants as well as protists, and number as many as 50,000 per square meter (dry weight of only 36 milligrams). Here too are found more than 29

Collembola and a pin head. *"The springtails are under the moss as well as in clumps of soil. They crawl and occasionally hop like fleas."*

species of nematodes, which number from one to five million per square meter and which add 60 milligrams to the dry weight biomass. They eat living plant and protist cells, and are themselves preyed upon by several species of predatory fungi.

The tardigrades of Antarctica have not been well studied. These minute animals typically have four pairs of short legs with claws and a suctoral pharynx. They feed on living moss cells and plant detritus, with some species also eating rotifers and nematodes. At least seventeen species are known from the Peninsula and the maritime antarctic islands. On Signy Island the population in the moister moss beds is less than 90,000 per square meter (160 milligrams dry weight).

Tardigrade

Arthropods

Although at least one species of insect (a chironomid midge—*Belgica antarctica*) is known from the Peninsula, the terrestrial community is dominated by two other sorts of arthropods—mites (Acari) and springtails. The mites are represented by about 32 species with densities as high as 200,000 per square meter (12 milligrams dry weight). Some of these are herbivores, eating algal cells gleaned from among the moss plants; while others are predaceous, attacking their peers as well as springtails.

Because of their size and ubiquity, the Collembola are probably the best-known group of terrestrial antarctic herbivores. About eighteen species of springtails are known from the Peninsula and its islands, reaching densities exceeding 21,000 (.22 grams) per square meter. The most common species here and throughout much of the Antarctic Peninsula is *Cryptopygus antarcticus*. While much is yet to be learned of the life history of this ubiquitous herbivore, its known biology is instructive of the adaptations that success in the Antarctic requires.

Cryptopygus antarcticus is the dominant terrestrial arthropod in Antarctica and its offshore islands south of the Antarctic Convergence. For that

The midge, *Belgica antarctica*

reason it would not be incorrect to refer to it as the "Antarctic Spring-tail," although a formal common name does not exist. It is classified in the superfamily Entomobryoidea (in common with most other Collembola in the Antarctic), and family Isotomidae. There are about twenty additional species of springtails in the genus *Cryptopygus,* all of which are restricted to the Southern Hemisphere. The antarctic springtail, however, does not range north of the Antarctic Convergence except at Macquarie Island in the Pacific Sector. In the Antarctic it is recorded from all terrestrial habitats free from permanent ice and snow. It comprises 90 to 100 percent of the arthropods found in bare rock and gravel habitats near penguin rookeries; 5 to 20 percent of arthropods inhabiting patches of lichens; and 70 to 90 percent of the dense fauna living in the moss turf.

This little springtail grows to a maximum body length of about two millimeters, and as an adult is a shiny, iridescent black. Its paler younger stages grow though a series of molts of undetermined number to reach maturity. These springtails are usually found in the summer in the upper three centimeters of the substrate. Other Collembola and some mites burrow deeper than this, particularly in the porous fabric of moss turf. In the winter most of the arthropod community dig somewhat deeper.

The antarctic springtail appears to be gregarious. Certainly its horizontal distribution is clumped, and aggregations are the norm. Indeed, masses of eggs that number in the hundreds are sometimes found. Because a single female can lay only a few (three to twenty) eggs at a time, it appears that springtails aggregate at least for the purpose of egg laying.

Because the eggs are laid in masses, the distribution of newly hatched springtails is also clumped, and large numbers of these tiny animals seem to appear as if by magic during brief bouts of favorable weather. Densities of about thirty animals per square centimeter are not unusual.

All life stages of the antarctic springtail can be collected at any season of the year. Mating and egg laying apparently occur whenever temperatures rise above the freezing mark. Considering the harshness of the environment, this opportunism is a necessity for survival in the Antarctic.

Information on the incubation time of springtail eggs is scanty. Clutches brought into the lab and held at a constant temperature well above freezing hatched at times up to ninety days later. It appears that eggs laid in one summer season may not hatch until the next. Adults held in the lab at 7°C (high by antarctic standards) survived for more than twelve months. In their normal environment antarctic springtails may live for a number of years. Long lifetimes are not unusual among the marine birds and mammals of the Antarctic, and this feature extends even to tiny herbivores.

The antarctic springtail feeds on the hyphae and spores of fungi and, principally, upon green algal cells that grow in the habitats it favors. As

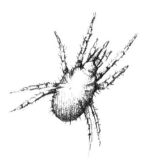

A typical mite

The Antarctic Springtail

Mass of springtail eggs

a primary consumer it harvests only a small portion of the sunlight that was converted into the organic matter of the algal population. Having consumed this energy, a fraction of it is converted into the substance of the springtail. Most is spent, however, on the metabolic activity that allows these tiny creatures to make the prodigious leaps that characterize the group to which they belong. Energetic analyses of moss turf and moss carpet communities at Signy Island show that the invertebrate and protist fauna consume less than 0.04 percent of the net annual production of mosses. Most of the food base consists of green algae, microorganisms, and detritus. Altogether something between 50 and 76 percent of the total net primary production is eaten.

FRESHWATER LAKE HERBIVORES

Of course the biological production on the land areas of the Antarctic is not confined to terrestrial forms. Lakes occur on the Peninsula, its islands, and places in East Antarctica. Several lakes on Signy Island are the best studied and present a picture of productivity quite different from what one might expect.

Although most of the freshwater lakes can be classed as oligotrophic on the basis of their low nutrient concentrations and biological productivity, some are very much enriched due to their proximity to bird colonies or their use as seal wallows. The diversity of algae and phytoplankton in such lakes is high, and their biological productivities can be an order of magnitude above that of similar lakes in the Arctic. This is reflected in populations of herbivores that include copepods, cladocerans, and rotifers. Yet the number of lakes that have been discovered in the Antarctic is not large (55 or so), and their total surface area (and energy production) is minuscule compared to the vastness of the continent.

Because of the relatively small area of productive terrestrial and freshwater habitat in the Antarctic, the energy which the herbivores transmit to their predators is certainly an incredibly tiny part of the solar radiation that falls on the entire region. To follow a more significant fraction of that energy we must turn to the sea, where the major percentage of Antarctica's fuel is converted and combusted.

MARINE HERBIVORES

It is in the vast Southern Ocean, all 36 million square kilometer's worth, that most of the Antarctic's primary production goes on. The major planktonic photosynthetic organisms consist of diatoms and, to a lesser extent, dinoflagellates. Many of these phytoplankters are very small. The herbivores that eat them tend themselves to be small. Some of the tinier herbivorous elements in the zooplankton include protists such as tintinnids (ciliates), radiolarans, foraminifera, and the younger life stages (nauplii) of copepods. Many of these organisms feed on diatoms and dinoflagellates that are in the nannoplankton size range, having diameters of less than 27 micrometers. Many of the cells that they eat are only 3 to 12 micrometers across. Because of the difficulties of studying these exceedingly small elements in the plankton, they are poorly known in most seas, and especially so in the Southern Ocean. Much of the usual oceanic sampling in the Antarctic uses plankton nets with mesh so coarse that only the larger sizes of copepods and krill are routinely taken.

There is a tendency to consider the adults of antarctic copepods and euphausiids (krill) as primary herbivores—the direct consumers of diatoms and dinoflagellates. In fact the adults of these zooplankters probably eat considerable amounts of tintinnids, copepod nauplii, forams, and radiolarans, and the young stages of krill. Food webs are seldom simple!

The business of being a herbivore in the Southern Ocean is different in some respects from that position in other oceans. Elsewhere the production of diatoms and dinoflagellates is often conditioned by the availability of nutrients that typically stimulate pulses of growth in the spring and fall. In the Antarctic nutrients are seldom lacking. There is but a single pulse that results from the long days of sun in the summer with its attendant ice melt.

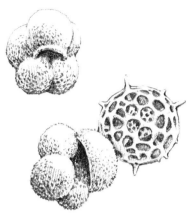

Radiolaran and foraminifera

Copepods, krill, and other members of the zooplankton community must seek some succor to get them through the dark and lean times of the ice-bound winter's long night. For some, this means belt-tightening. For others it means a shift in diet from plants to zooplankton or bacteria. Some take sustenance in the epontic community immediately under the ice. In one way or another all adapt to this unique challenge.

To discuss adaptations in the herbivorous zooplankton of the Southern Ocean it will serve our purpose to break this system down into three parts—the near-shore and ice communities (neritic and epontic communities), the copepods, and the krill.

Near-Shore and Under-Ice Communities

For good logistical reasons the near-shore (neritic) and under-ice (epontic) pelagic communities in the Antarctic are not well studied. The areas nearest shore are most likely to be ice-covered and, like water under the pack ice, are difficult to access. There is a thriving cryopelagic community based on diatoms and dinoflagellates frozen directly into the ice and appended to its undersurface. These primary producers are able to live on the diminished light levels that are transmitted through snow-covered ice. The nutrient chemicals that are concentrated in the brine salted out of

Fast ice hugging the shore

forming frazil are important factors in their growth and reproduction. The concentrations of diatoms are greatest right at the bottom surface of the ice. Here, in addition to the phytoplankters, are a host of zooplankton that gather to graze on the rich "grass" embedded in and pendant from the undersurface of the ice. Prominent among these animals are numerous copepods, amphipods, mysids (another group of shrimp-like crustaceans), and euphausiids. Although all of these crustaceans are mainly filter feeders, aquarium-held individuals actually gnaw away at phytoplankton embedded in chunks of ice presented to them.

These most southern populations of crustaceans all share characteristics that hallmark life in the Southern Ocean. For example, growth rates vary according to season, being highest when food is most abundant. In the epontic community this can be in the winter and spring months when food concentrates at the ice-water interface. Consonant with slow and variable growth rates, longevity is typically extended in many of these species. In the case of the neritic mysid, *Antarctomysis maxima,* populations studied in the South Orkney Islands have a generation time (four years) that is twice as long as that found in a population of the same species not too far to the north at South Georgia.

These zooplankton easily switch from one sort of food source to another. Most are opportunistic feeders that are herbivores at some times and carnivores or detritivores at others. A key to survival in an environment that is predictable only in its harshness seems to be this kind of opportunism. Physiologically specialized for a cold, dimly lit environment, these animals are nonetheless feeding generalists.

Even though life under ice does not strike us as comfortable, there are advantages that speak well for this existence. The temperature, though cold ($-1.84°C$), is constant. The ice cover and latent heat of fusion of water effectively buffer temperature changes in the air above so that organisms living in the water can be certain of the temperature. The ice also protects the water and its organisms from wind and waves. Although wind waves normally mix the sea, distributing dissolved gases and nutrients, waves also threaten to submerge phytoplankton and their predators from the sunlit surface stratum. Ice provides a haven of stability in this otherwise tumultuous environment.

Although most of the epontic community nestles close against the underside of the ice cover, a particular pelagic community exists well below. Foraminifera, radiolarans, tintinnids, and other protists move in the depths. Most of these are found in the upper 100 meters of the water column, and constitute a definite, if relatively small, component of the epontic community.

PELAGIC COMMUNITY

At its maximum, with about 20 million square kilometers of the Southern Ocean covered by sea ice, the epontic community is an extensive factor in the economy of the Antarctic. However, when the days lengthen in the spring and early summer and the ice melts, its cryopelagic organisms are freed to seed the open sea. It is then that the productivity burst for which the Southern Ocean is famous takes place. This is quickly reflected in the dynamics of the copepod and krill populations.

Sea ice melting and dispersing—
scattered growlers and bergy bits

Typical copepod

Copepods

In temperate and tropical oceans copepods make up one of the most conspicuous and important elements in the zooplankton. Many species are herbivores, although broad omnivory appears to be the norm. These little crustaceans feed by raking the water with bristled appendages such as their second maxillae. These act like tiny paddles that push parcels of water containing food into the mouth. Rather than being "nonselective" feeders, many copepods seem to sense their food chemically or by mechanical vibrations, are attracted to it, and carefully sort through it.

Once food particles such as diatoms are swallowed into the gut, peculiar "peritrophic membranes" are secreted to enclose a bolus of food in a sort of sausage casing while digestion and assimilation take place. The exact benefits of the peritrophic membrane are uncertain. It surely costs the copepod some energy to fabricate, and probably provides the animal with mechanical protection from the tiny slivers of silica that result from the crushing of diatom frustules during ingestion. Animals that eat glass deserve some inner protection from the shards! The peritrophic membrane may also encase bacteria and other potentially damaging microorganisms that are swallowed with the food.

If the benefits of the peritrophic membrane are conjectural, one environmental effect is not. When digestion is completed the fecal material is excreted by the copepod still packaged in this membrane. Copepods thus produce discrete fecal pellets that are, in effect, little torpedoes having sinking rates much faster than the flocculant, unconsolidated wastes of other animals. This results in rapid downward transport of the considerable amounts of nutrients that reside in the pellets. These nutrients are released into the water column only after bacterial degradation of the membrane occurs. And this may happen at depths well below the surface mixing zone.

In shallow seas copepod fecal pellets accumulate on the bottom as a significant source of sediments. In deeper waters the nutrients can be released above the bottom, but in water much too deep to be recycled immediately through the phytoplankton. Indeed, the nutrient richness of the Warm Deep water flowing into the Southern Ocean comes, in part, from accumulated copepod fecal material raining through it during its slow course southward in the Atlantic, Pacific, and Indian Oceans. The upwelling and general instability of the Southern Ocean allows its own copepod detritus to be recycled more locally than in the better stratified oceans of the world.

Each adult copepod begins life as a much smaller nauplius larva. After six molts through ever larger nauplius stages, the copepod enters the copepodite form and passes through another series of molts to arrive finally at the adult stage. In the Southern Ocean the nauplii are relatively unstudied, and a fair amount of information exists only on the larger copepodite and adult stages.

Copepods characteristically show daily episodes of vertical migration. Although each species (and sometimes specific life stages of each species) has its preferred depth, all but the deepest make migrations toward the surface during the early evening hours; and then retreat with the onset of daylight. The purpose of these nocturnal forays toward the surface could be to feed in shallower water (where the density of phytoplankters is high) by night, and to hide from predators in the darker depths by day. This pattern of daily vertical migration is shown by many of the Southern Ocean copepods. More impressive, however, are the vertical migrations which many of these same copepods make on a seasonal basis.

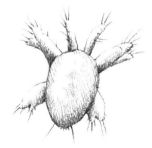

Copepod nauplius

More than 100 species of copepods are known from the Southern Ocean. More than a third of these are abyssal forms that reside in the deepest layers of the ocean, existing on diets that include detritus, bacteria, and zooplankton. In the surface waters two of the most common copepods are *Rhincalanus gigas* and *Calanoides acutus.* Together they form a large percentage of the copepod biomass in the upper layers of the ocean. They are large species with adult females measuring more than 5 millimeters in length, and are generally restricted to waters south of the Antarctic Convergence. In the far south (the Weddell Sea, for example) *R. gigas* is rare. Both species show pronounced seasonal vertical migrations that serve to transport them north and south, through the length of the Southern Ocean. In the austral fall, large populations of the young copepodite stages of these species occur in the north, just below the convergence. At this time of year the copepodites sink from the surface into the Warm Deep water that carries them south in relative comfort during the winter. In the spring the maturing copepods spawn as they rise to the surface waters at the edge of the pack ice. The males of *C. acutus* die soon after spawning and are rare in summer surface collections. The nauplii, copepodite, and remaining mature stages then drift north in the Antarctic Surface water during the summer, until fall finds them once again at the northern limit of their range. The animals sink again and the cycle repeats itself. In the winter two thirds of the zooplankton biomass is found below 200 meters. In the summer most of this swims in the surface layers.

At the highest latitudes of the Southern Ocean the large species of copepods are rare. Here instead are smaller kinds such as *Ctenocalanus vanus* and *Oithona* sp. Only 10 percent of the total summer population consists of adults. In these waters, which are subject to extensive winter ice cover, the biomass of the copepods is much smaller than in the more open regions of the ocean, although swarms as dense as 60,000 copepods per cubic meter have been found at the undersurface of fast ice.

In general the standing crops of herbivorous zooplankton in the Southern Ocean is thought to average more than 100 milligrams per cubic meter in the surface layer. Perhaps half of this total is made up of copepods, plus some tunicates, arrowworms, amphipods, fish larvae, and others. The remaining 50 percent consists of a single species of euphausiid, the krill, *Euphausia superba*.

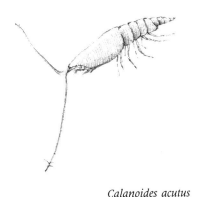

Calanoides acutus

Krill

It is difficult to overestimate the importance of krill to the Southern Ocean. These crustaceans formed the dietary staple of the bygone stocks of baleen whales, and today support burgeoning populations of seals, penguins, fish, and squid. As a keystone species in the flow of energy through the entire antarctic region, it is important to look closely at the biology of krill.

Antarctic krill (Euphausiacea) belong to the same superorder (Eucarida) of crustaceans as do the more familiar Decapoda that include the crabs, lobsters, and shrimps. It is something of a zoogeographic irony that, although euphausiids are so abundant and important here, the decapods are poorly represented in the Southern Ocean—the only major ocean from which they are almost missing.

There are 85 species of euphausiids found in the world, nearly all being

Euphausia superba

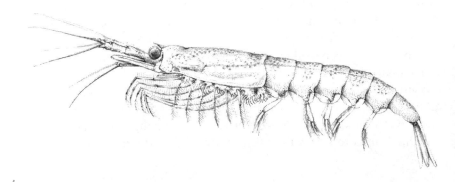

associated with the epipelagic habitat of open ocean waters. To an un-critical eye, these animals appear as tiny shrimp or prawns. An important difference, however, is the presence of thoracic legs which are not modified for walking on the bottom, or equipped with chelicerae (pincers) for feeding or defense. There are also technical differences in the shape of the mouth parts. Most euphausiids, including *Euphausia superba,* shine with photophores, some of which are found lining the lateral edges of each abdominal segment. In the ocean these seem sensitive to the amount of dissolved oxygen in the blood. They normally glow constantly, but increases in blood flow to the abdomen that occur during brief periods of burst swimming cause the photophores to produce flashes of bioluminescence.

While the greatest numbers of species of euphausiids occur in tropical and subtropical seas, a few species dominate in waters of high latitudes, particularly in the Southern Ocean. Here live 12 species that include the neritic *Euphausia crystallorophias,* which is prominent in the epontic, under-ice community. *Euphausia superba* is also found under sea ice, but reaches its greatest abundance in open areas of the Southern Ocean, particularly in the gyres emanating from the East Wind Drift, in straits near the Peninsula, and near areas of upwelling associated with the Antarctic Divergence.

E. superba is one of the larger species of krill, reaching a final adult length of about 60 millimeters (about 2.5 inches) or a little more; with individual weights somewhat in excess of a gram. The adults are rich in carotenoid pigments that give dense swarms of these animals a reddish cast in the water.

The life of *E. superba* begins when eggs, fertilized by sperm stored in the female's spermatheca, are released to the water column. A single female spawns several batches of 500 to 8,000 eggs in a summer season—as many as 20,000 eggs all told. The eggs sink at rates of as much as 200 meters per day to depths of more than a thousand meters where they hatch as tiny nauplii larvae that live on yolky material stored in the eggs. Each nauplius develops into a stage known as the metanauplius, then to a calyptopis, furcilia, and finally the adult form. It takes about four months to develop from the egg to the last furcilia at 0°C in the laboratory. The total life cycle in *E. superba* takes six to seven years to complete. The mouth parts become functional in the calyptopis, and by then the larval krill have moved to the surface layers where they become active feeders. The furcilia are active herbivores and vertically migrate to near the surface at night. In the adult form, from about 40 millimeters in length, krill are capable of eating not just diatoms and dinoflagellates, but a wide variety of foodstuffs including protists, copepods, and even their own eggs and larvae. Recent studies suggest that adult krill live as long as five years in that stage, regressing in size when food becomes scarce.

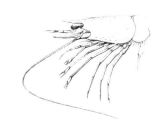

Thoracic legs of krill (above), and shrimp (below)

Eggs, nauplius, metanauplius, and calyptopis stages (top to bottom) of krill

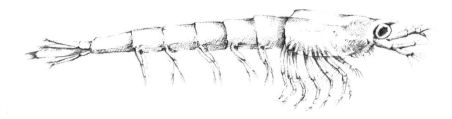

A krill furcilia

Episodic feeding is accomplished by the rapid wheeling motion of the six pairs of thoracic legs that are lined with numerous hair-like setae. This feeding behavior is stimulated by a wide variety of water-borne chemical cues perceived by the krill. Water currents set up by the legs carry a stream of food particles that are concentrated in the "food basket" dorsal to and between the legs. As much as 450 milliliters of water is filtered per hour during bouts of feeding. At intervals the krill swallows the accumulated material—some say indiscriminately, others report only after sorting out inedible particles. Although a variety of prey may be eaten, certainly diatoms provide the bulk of the food of krill, particularly in the sunlit summer surface waters.

Like copepods, krill ensheathe their ingested food in peritrophic membranes that result in the formation of discrete fecal pellets. These, like those of copepods, have sinking rates that could remove nutrients from surface waters on a large scale. That potential is reduced in the Southern Ocean both by the significant upwelling and turbulence, and by the life cycles of the copepods and krill that live there. The younger stages (nauplii and metanauplii) reside during the winter mainly in the Warm Deep water mass, well below the surface layer. Here they may be able to absorb nutrients (dissolved organic molecules, vitamins, and salts) directly from the water without active feeding. In their vertical ascent they serve to bring these nutrients once again to the surface where recycling can occur.

A critical feature of euphausiids in general, and *E. superba* in particular, is their swarming behavior. Immense groups of krill gather as discrete shoals, sometimes many kilometers in diameter. Within the larger shoal are "schools"—groups of individuals that are oriented to each other in a parallel fashion. Also observed are "swarms" that are extremely dense aggregations of krill, not so obviously organized. *E. superba* swarms have been reported with densities approaching 60,000 individuals per cubic meter of water, although values half of this figure are more usual. This means biomass densities as high as 60 kilograms per cubic meter. Direct measurements of biomasses of 35 kilograms per cubic meter are on record.

The swarming behavior of krill undoubtedly serves a reproductive function, facilitating the location of mature females by males. It also may

allow krill to crop most effectively the phytoplankters which themselves are distributed in patches along the gyres and ocean fronts of mixing. Finally the schooling behavior of krill allows them to communicate visually the location of predators. Schools of krill, frightened by human divers investigating them, are reported to suddenly molt their old exoskeletons en masse and flee, leaving behind just their cast shells that might act as decoys to confuse predators.

Considering an average zooplankton biomass of about 105 milligrams per cubic meter in the surface layer of the Southern Ocean, the presence of swarms of krill more than four orders of magnitude denser suggests a "patchiness" to this ocean's resources that has several untoward consequences for the krill. In the first instance the presence of several kilograms of krill in a small volume of sea water augurs the consumption of oxygen beyond the environment's ability to provide the levels needed to satisfy the full metabolic demands of these crustaceans. Oxygen consumption rates in krill are reported to range from 0.49 to 1.88 microliters of oxygen per gram (wet weight) per hour. They require high oxygen partial pressures. At less than about 80 percent of saturation significant mortality occurs. One can roughly calculate that two kilograms of krill would extract enough oxygen from one cubic meter of sea water at 0°C to get them in trouble within an hour. Dense swarms of krill must therefore be on the move, both to find food and to keep them in adequate oxygen. Swarms are usually configured in narrow bands and sheets that give each krill ready access to well-oxygenated water.

Euphausiids swim at cruising speeds by paddling their abdominal legs (pleopods). The synchronized beating of these legs propels the animals as fast as eight times the body length per second. In emergencies they rapidly flex the abdomens to shoot backwards at burst velocities of about eleven body lengths per second. *E. superba* orients visually to light, maintaining a position with the long body axis at right angles to the direction of the incident light. At night and at depth the photophores probably allow the krill to maintain mutual orientation so that shoals, with their included schools and swarms, stay intact.

In addition to flirting with physiological distress, dense aggregations of krill invite significant predation and exploitation. Swarms of krill were the mainstay of the great herds of baleen whales in the Antarctic. Today these swarms are preyed upon by a host of birds (petrels, penguins), squid, and mammals such as seals, sea lions, and the few remaining whales. Despite this predation, enormous numbers of krill remain.

Estimates of the standing crop of krill in the Southern Ocean vary widely, but the median runs close to between 500 and 750 million metric tons, with an annual production of between 750 and 1,350 million metric tons. Because this huge biomass is concentrated by the swarming habit of krill, they are accessible not only to whales, seals, and penguins, but to modern fishing fleets as well. Exploitation of this resource has only begun, but some fishery experts feel that the krill catch may someday exceed the present world-wide harvest of all seafood. The politics and problems associated with this offing will be considered in chapter 10. Meanwhile, suffice it to say that *Euphausia superba* is in every way the central block in the ecology of the Antarctic. In the next chapter we will begin to explore the elements of this ecosystem that heavily depend on these small crustaceans.

ADDITIONAL READING

Abele, L. G. 1982. Biogeography. In L. G. Abele, ed., *The Biology of Crustacea,* 1:242–293. New York: Academic Press.

Block, W. 1984. Terrestrial microbiology, invertebrates and ecosystems. In R. M. Laws, ed., *Antarctic Ecology,* 1:163–236. London and Orlando, Fl.: Academic Press.

Boyd, C. M., M. Heyraud, and C. N. Boyd. 1984. Feeding of the antarctic krill, *Euphausia superba. Journal of Crustacean Biology,* 4(Spec. No. 1):123–141.

Davis, R. C. 1981. Structure and function of two antarctic moss communities. *Ecological Monographs,* 51(2):125–143.

Ettershank, G. 1985. Population age structure in males and juveniles of the antarctic krill, *Euphausia superba* Dana. *Polar Biology,* 4:199–201.

Greenslade, P. and K. A. J. Wise. 1984. Additions to the collembolan fauna of the Antarctic. *Transactions of the Royal Society of South Australia,* 108(4):203–205.

Hamner, W. M. 1984. Aspects of schooling in *Euphausia superba. Journal of Crustacean Biology,* 4(Spec. No. 1):67–74.

Hamner, W. M., P. P. Hamner, S. W. Strand, and R. W. Gilmer. 1983. Behavior of antarctic krill, *Euphausia superba:* Chemoreception, feeding, schooling, and molting. *Science,* 220:433–435.

Heywood, R. B. 1984. Antarctic inland waters. In R. M. Laws, ed., *Antarctic Ecology*, 1:279–344. London and Orlando, Fl.: Academic Press.

Ikeda, T. 1985. Metabolic rate and elemental composition (C and N) of embryos and nonfeeding early larval stages of antarctic krill (*Euphausia superba* Dana). *Journal of Experimental Marine Biology and Ecology*, 90:119–127.

Ikeda, T. and P. Dixon. 1984. The influence of feeding on the metabolic activity of antarctic krill (*E. superba* Dana). *Polar Biology*, 3:1–9.

Ishii, H., M. Omori, and M. Murano. 1985. Feeding behavior of the antarctic krill, *Euphausia superba* Dana. I. Reaction to size and concentration of food particles. *Transactions of the Tokyo University of Fisheries*, No. 6, pp. 117–124.

Janetshek, H. 1970. Environments and ecology of terrestrial arthropods in the high Antarctic. In M. W. Holdgate, ed., *Antarctic Ecology*, 2:871–885. New York: Academic Press.

Knox, G. A. 1970. Antarctic marine ecosystems. In M. W. Holdgate, ed., *Antarctic Ecology*, 1:69–96. New York: Academic Press.

Laws, R. M. 1984. The ecology of the Southern Ocean. *American Scientist*, 73:26–40.

Mauchline, J. 1980. The biology of mysids and euphausiids. *Advances in Marine Biology*, 18:1–677.

Mauchline, J. and L. R. Fisher. 1969. The biology of euphausiids. *Advances in Marine Biology*, 7:1–454.

Morely, J. J. and J. C. Stepien. 1984. Siliceous microfauna in waters beneath antarctic sea ice. *Marine Ecology Progress Series*, 19:207–210.

Ross, R. M. and L. B. Quentin. 1985. The effect of pressure on the sinking rates of the embryos of the Antarctic krill, *Euphausia superba*. *Deep-Sea Research*, 32(7):799–807.

Schenker, R. 1984. Effects of temperature acclimation on cold-hardiness of Antarctic microarthropods. *Revue d'Écologie et de Biologie du Sol*, 21(2):205–220.

Smith, H. G. 1984. Protozoa of Signy Island fellfields. *British Antarctic Survey Bulletin*, 64:55–61.

Smith, H. G. 1985. The colonization of volcanic tephra on Deception Island by protozoa: Long term trends. *British Antarctic Survey Bulletin*, 66:19–33.

Tanimura, A., T. Minoda, M. Fukuchi, T. Hoshiai, and H. Ohtsuka. 1984. Swarm of *Paralabidocera antarctica* (Calanoida, Copepoda) under sea ice near Syowa Station, Antarctica. *Antarctic Record*, 82:12–19.

Tilbrook, P. J. 1970. The terrestrial environment and invertebrate fauna of the maritime Antarctic. In M. W. Holdgate, ed., *Antarctic Ecology*, 2:886–896. New York: Academic Press.

Tilbrook, P. J. 1970. The biology of *Cryptopygus antarcticus*. In M. W. Holdgate, ed., *Antarctic Ecology*, 2:871–885. New York: Academic Press.

Turner, J. T. 1984. The feeding ecology of some zooplankters that are important prey items of larval fish. *NOAA Technical Report NMFS*, 7:1–27.

Usher, M. B. and R. G. Booth. 1984. Arthropod communities in a maritime antarctic moss-turf habitat: Three dimensional distribution of mites and Collembola. *Journal of Animal Ecology*, 53:427–441.

Vervoort, W. 1965. Notes on the biogeography and ecology of free-living marine copepoda. In P. van Oye and J. van Mieghem, eds., *Biogeography and Ecology in Antarctica*, pp. 381–400. The Hague: Junk.

Voronina, N. M. 1970. Seasonal cycles of some common copepod species. In M. W. Holdgate, ed., *Antarctic Ecology*, 2:162–172. New York: Academic Press.

Ward, P. 1984. Aspects of the biology of *Antarctomysis maxima* (Crustacea: Mysidacea). *Polar Biology*, 3:85–92.

CHAPTER 5

SECONDARY CONSUMERS— FISHES, SQUID, THE BENTHOS

FISHES

More than twenty thousand species of fishes live in the world today. These finny vertebrates dominate by numbers of individuals as well as species. More kinds of fishes exist than all other vertebrates put together. Among them about 58 percent (11,600) are primary marine forms, of which 7,980 are found in shallow warm seas. With about 10 percent of the world's oceanic acreage the Southern Ocean should harbor a proportionate number of fish species. That is, 10 percent of the marine forms (1,160 species), or at least 10 percent of the 3,400 species that are confined to shallow-cold, deep pelagic, and deep-benthic habitats. Remarkably, no more than 120 species of fish are recorded from waters south of the Antarctic Convergence.

The meager fish community in the Southern Ocean is strikingly different from fish assemblages found in Arctic waters. Prominent among the

A typical nototheniiform fish, *Notothenia gibberifrons*

northern fish communities are abundant schools of pelagic fishes that feed on zooplankton and thus constitute a major link in arctic and boreal oceanic food chains. Fishes like herring *(Clupea)*, capelin *(Mallotus)*, and sand lance *(Ammodytes)* form enormous schools in northern waters; and themselves become the principal food of sea birds, seals, and whales. These sorts of fishes are absent in the Antarctic fauna, even though they seem well-adapted to survive in the cold and rough conditions of the Southern Ocean.

One factor responsible for the dearth of pelagic, schooling southern fishes is the comparatively small area and great depth of the antarctic continental shelf. In the north, herring, capelin, and sand lance spawn in shallow waters where their fertilized eggs develop at or near the bottom. This shallow spawning habitat is virtually unknown in the Antarctic.

In the Southern Ocean, almost 75 percent of the fish genera, and 90 percent of its species are endemic. That is, they occur nowhere else in the world. Moreover, many of these fishes show physiological adaptations that are extreme, even for such a diverse group of animals as bony fishes.

Seventy-five percent of the Antarctic fish fauna consists of one order of mainly benthic fishes, the Nototheniiformes. The order contains but five families, four of which are found south of the Antarctic Convergence. Most of these sculpin-like fishes are spiny, with two or three lateral lines, only a single nostril on each side, and are technically characterized by the bony structures of their pectoral fin supports. Few nototheniiform species grow to more than 50 centimeters in length, and, while the young of several kinds live pelagically in the open ocean, only one species, the

Nototheniops nudifrons

Antarctic silverfish *(Pleurogramma anatarcticum),* is well adapted to operate as a year-round, pelagic planktivore.

The largest and most typical nototheniiform family, the Nototheniidae, includes at least 36 species recorded from south of the Antarctic Convergence. Of these, 15 species live in the waters off the Antarctic Peninsula, while 10 others occur off East Antarctica. One of the most distinctive species (certainly for its size) is *Dissostichus mawsoni,* a fish that grows to a meter and a half in length and weighs over 25 kilograms (an unsubstantiated record of an individual of 64 kilograms exists). This is the largest fish known from the Antarctic and dwells in the deep pelagic zone of the Ross Sea. More typical are small species such as *Pagothenia bernacchii, P. borchgrevinki,* and *Nototheniops nudifrons.* These fishes hug the bottom as adults, grow to mature sizes of not much more than 35 centimeters, and eat a variety of food items ranging from krill through bottom amphipods, isopods, algae, mysids, and echinoderms. Growth is typically slow, maturity occurring at about five years. Many species live from ten to fifteen years.

Nototheniiform fishes are not fecund. Females ripen only a few hundred to several thousand eggs per reproductive season. By contrast, an Atlantic cod off Newfoundland produces as any as six million eggs per year. Nototheniiform eggs are large, with diameters up to 4.8 millimeters in *Notothenia rossii marmorata.* The size of the eggs (and the amount of food they store) increases with latitude. Fishes caught near the Peninsula have yolkier eggs than those of the same species further north off South Georgia. Many nototheniids spawn in early winter so that the juveniles are large enough the following summer to crop plankton during the summer bloom. The time of spawning is keyed to larval survival rather than to the time of the year when the adults are in the best condition.

The large eggs are deposited on the bottom where nest defense and

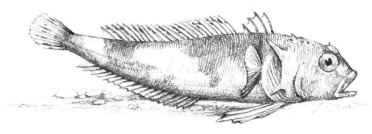

Harpagifer, a plunderfish

brood protection seem to occur in most species. Extended care of young is a recurrent reproductive theme among animals living in extreme environments. In the Antarctic groups as diverse as echinoderms, mollusks, and crustaceans practice parental care. It should not surprise to find that fishes also look after their young.

Other Antarctic nototheniiforms include 14 or 15 species of plunder fishes of the family Harpagiferidae, 15 species of Antarctic dragon fishes (family Bathydraconidae), and about 15 species of ice fishes (family Channichthyidae). The plunder fishes are small (10 to 30 centimeters), scaleless fishes each with a characteristic barbel on the lower jaw. Most live near the bottom deep on the continental shelf, although a species of the genus *Harpagifer* is common in tide pools on South Georgia, and is found in shallow water at the northern tip of the Peninsula.

The Antarctic dragon fishes are elongate animals with pike-like snouts and no first dorsal fin. Some grow to fifty centimeters, and one, *Prionadraco evansi,* is semitransparent. Its blood vessels, gills, and vertebral column can be seen though its skin. Most of these fish are caught near the bottom in the deeper water on the edge of the Continental Shelf at depths of 500 to 700 meters. One species, *Bathydraco scotiae,* is the deepest dwelling nototheniiform fish, having been caught as deep as 2,579 meters.

Ice Fishes

One of the most unusual groups of fishes (or any other group of vertebrates) is the family Channichthyidae (ice fishes). These include about 15 Antarctic species, at least four of which are found off the Peninsula. They are caught at depths of 100 to 700 meters. Ice fishes lack scales and have large, spatulate heads with strong teeth arming their large jaws. They are predaceous and average larger in body size than most other antarctic fishes, with some species growing to about 67 centimeters long. The real distinction of the ice fishes was not widely recognized until 1954 when it was learned that they are the only known adult vertebrates to lack completely the respiratory blood pigment, hemoglobin! Channichthyids thus have colorless blood, reduced spleens, and large cream-colored gills—in contrast to the pink or red gills of other fishes. The ice fishes were originally

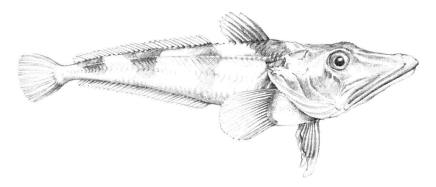

Chaenocephalus—an ice fish

reported to lack red blood cells (erythrocytes). Subsequent studies demonstrated, however, that they have a fraction (¹/₁₀ to ¹/₉₀) of the erythrocyte numbers of normal fishes. It's just the hemoglobin that is missing.

Hemoglobin normally binds and transports oxygen molecules from the point where they enter the blood (in the gill or lung) to the tissues, where metabolizing cells require oxygen to complete the reaction sequence that liberates energy from food molecules. Because oxygen is poorly soluble in aqueous solutions (including blood plasma and sea water) the presence of hemoglobin allows blood to transport for more oxygen than it could carry in simple solution. In most marine fishes 100 milliliters of blood plasma normally carry only about 0.3 to 0.5 milliliters of oxygen, with more active fish (mackerel, for example) having higher oxygen carrying capacities than sedentary fishes.

The total amount of blood of bony fishes is considerably lower than that of most vertebrates, being only in the neighborhood of 3 percent of body weight. A typical fish weighing one kilogram has only about 30 milliliters of blood, something less than half of which is fully oxygenated, arterial blood. The fish contains about 1.5 milliliters of oxygen at any given time, and delivers about 3 milliliters of oxygen to its tissues each time it circulates its blood volume (in a fish like a cod this is about a minute and a half). Without hemoglobin, the same fish would deliver only about 0.1 milliliters of oxygen in the same time. How can the Channichthydae, a group of actively foraging fish, do without hemoglobin? What are the advantages of this condition?

The water of the Southern Ocean is well oxygenated. Oxygen concentrations here are higher (*ca* 7 milliliters of oxygen per liter of sea water) and more predictable than in most other oceans of the world. Also the pace of life in these waters is slow. The chronically depressed, uniform sea temperatures (nearly always below 0° C) limit metabolic activity.

The physiology of ice fishes is so closely adjusted to low temperatures that their systems fail if the environmental temperatures are raised just a few degrees in the laboratory. Measurements of muscle activity in antarctic fish show that at 0°C they develop greater tensions than do cod at 8°C or marlin at 20°C. Despite their thermal adaptations, the measured respiratory rates of ice fishes are well below those of other fishes from warmer climes; and their metabolism is only half the average for their nototheniid peers.

Other Antarctic fishes have red blood cell counts and hemoglobin concentrations only half those found in temperate teleosts. Ice fishes complete a trend toward blood cell depletion.

At sub-zero temperatures, some fishes can exist anaerobically, needing no oxygen at all. Hemoglobin in fishes may function as an oxygen reservoir, rather than as a necessary transporter of oxygen. If the oxygen demand is low enough, and the environmental supply of oxygen is sure, it might cost the animal more energy to make and maintain hemoglobin than it benefits from having it. The production of red blood cells, and the hemoglobin with which to fill them, is an expensive process. The energy required for this is apparently not worth the effort in the Channichthyidae. The ice fishes rely instead on several less costly adaptations to satisfy their meager needs for oxygen.

First, the blood volume of ice fishes averages 2 to 3 times larger than that of other teleosts—as much as 9 percent of their body weight—providing more plasma in which to dissolve oxygen. Next, these fishes have a larger heart that expresses itself in a much higher cardiac output. The blood vessels are larger and blood is circulated more rapidly in channichthyids. Their gills are larger than average and present extensive surface area across which oxygen can diffuse.

A final adaptation to extracting oxygen from sea water involves the skin. The naked, richly vascularized skin of ice fishes makes a likely route for the diffusion of oxygen into the animal.

The Channichthyidae include species *(Champsocephalus esox)* that range north to Patagonia and the Falklands. The Antarctic forms make active forays to the oceanic pelagic zone where they eat krill. Some species grow faster, reach maturity quicker, and attain larger adult sizes than other nototheniids. Clearly, the loss of hemoglobin may be an advantage rather than the detriment that normal intuition suggests.

While nototheniiform fishes comprise the bulk of a rather sparse Antarctic fish fauna, other sorts of fishes are also found here. Eel pouts (family Zoarcidae) and sea snails (Liparidae)—fishes that are especially well-represented in Arctic waters—are also known from the Southern Ocean. Both of these bottom-loving groups live in deep shelf and slope water; and both are known for their brooding of eggs and parental care of young. Other fishes recorded in the Antarctic include two species of eel cods (Muraenolepidae)—the young of which are pelagic—and several abyssal forms like brotulids and grenadiers of the family Macrouridae. One unique species of flounder has been described from this region, as have two or three species of skates—the only elasmobranchs definitely known to penetrate south of the Antarctic Convergence.

Blood Antifreeze

All fishes in the Southern Ocean, and particularly those in the colder regions of the continental shelf, have a special problem posed by the cold ocean temperatures. Sea water freezes at about $-1.9°C$. Because bony fishes have blood and tissue fluid solute concentrations less than sea water, their bodies should freeze solid at temperatures of only $-0.7°C$. The options available to them in the face of these conditions are several. First, they could do nothing—allowing their body fluids to freeze when temperatures drop below $-0.7°C$. Frozen tissues, however, mean death for most fishes. A second, and viable, option is to avoid cold temperatures, seeking instead warmer layers of water. While temperature-directed migrations of fishes are not prominent features of the Antarctic—as they are in temperate oceans—some species do move to escape lethally cold conditions. For instance, *Notothenia kempi* shows none of the special physiological adaptations found in other Antarctic fishes. It avoids freezing by seeking warmer layers of water that intrude from the Warm Deep layer of the Southern Ocean. It is captured in winter well below the cold surface, in tongues of deep water as warm as $+1.0°C$.

Most Antarctic fishes adapt to subfreezing temperatures by stocking their blood with chemicals that both depress the freezing point and inhibit the formation of ice crystals below that point. The main molecules that accomplish these tasks are glycoproteins formed of repeating subunits that consist of a tripeptide with a disaccharide carried on it. These molecules work by interacting with the growing faces of incipient ice crystals, and interfere with ("poison") the addition of water molecules to the growing crystal lattice.

The effects of these glycoproteins are dramatic. Shallow water nototheniiform fishes like *Gymnodraco acuticeps* and *Pagothenia borchgrevinki*, which theoretically should freeze solid at temperatures around $-1°C$, have effective freezing points of $-2.48°C$ and $-2.75°C$ respectively. *Gymnodraco* lives in contact with anchor ice on the bottom of McMurdo Sound, and *P. borchgrevinki* actively forages in the cryopelagic community among platelets of frazil forming on the underside of sea ice.

Fishes living in deeper water have less dramatic antifreeze effects, in part because they never come in contact with ice crystals to seed their supercooled blood. Also the effects of pressure at great depth further depress the freezing points of their body fluids. The blood of the sea snail, *Paraliparis deVriesi*, for example, freezes experimentally at only $-1.15°C$, yet it lives at temperatures as low as $-2.1°C$ at 500 meters where sea water should freeze at $-2.28°C$. When these fish are raised to the surface in fish traps they freeze solid upon coming in contact with ice crystals.

The scarcity of pelagic fishes in the Southern Ocean is probably related to the deep and narrow continental shelf that offers little area for egg deposition. The production of floating or neutrally buoyant eggs by erstwhile pelagic species has not happened in the Antarctic. Such eggs (and the larvae that hatch from them) might be exposed to conditions too perilous (cold, turbulence, too little food) for survival through the winter. Remember, the eggs of pelagic krill and copepods sink to great depths before hatching.

All nototheniiform fish lack swim bladders, which in most pelagic fishes are used for buoyancy regulation. When the circumpolar current and Antarctic Convergence first isolated the continent about 25 million years ago, the resident fishes were most likely benthic forms that had previously lost their swim bladders. The few nototheniiform fishes that are pelagic in habit invest substantial energy in the synthesis and storage of fats and other lipids that, being less dense than water, provide positive buoyancy. *Pagothenia borchgrevinki Dissostichus mawsoni,* and *Pleurogramma antarcticum* are three such species. *P. borchgrevinki* forages in the cryopelagic community just beneath the ice surface; *Pleurogramma antarcticum* is pelagic in the surface waters around the continent; and *D. mawsoni* is a deep (300 to 500 meters) pelagic piscivore. All store far more lipid than bottom fish, mainly in pockets and connective tissue sacs under the skin, and in the body musculature. *Dissostichus* and the Antarctic silverfish are so fat that they are truly neutrally buoyant. In the case of the Antarctic silverfish, the energy used in creating and maintaining these fat stores means that they must have access to the immense stocks of krill living in the waters off the Peninsula.

Antarctic silverfish make up about 90 percent of the fish (by weight and number) collected in trawls sampling the surface and midwater pelagic zones off the Peninsula. These fish (which average only about 16 to 20 centimeters long) mainly eat copepods and krill, with *E. superba* being the most prevalent large prey.

The Antarctic silverfish is not the only krill eater among the fishes of the Southern Ocean. *Notothenia rossi* makes pelagic forays when adult, and *Nototheniops larseni* and *Champsocephalus gunnari* (an ice fish) seasonally rise to the pelagic zone to feed on euphausiids and copepods. These species have figured prominently in a commercial fishery that is developing near the Subantarctic islands of South Georgia and Iles Kerguelen. [(chapter 10).] Other croppers of krill include small numbers of southern hake *(Micromesisteius australis)* and Patagonian hake *(Merluccius hubbsi)* that migrate south from Argentinian waters to the Scotia Sea in the summer.

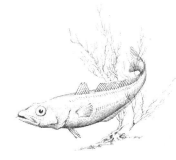

Antarctic silverfish

The huge standing crop of krill and other zooplankton is only slightly harvested by the depauperate fish fauna of the Antarctic. To follow the fate of more of these herbivores we must look to other groups of predators, including squid and communities of benthic invertebrates.

SQUID

Of the many biological mysteries in the Southern Ocean, one of the most interesting concerns the cephalopod mollusks we know as squid. These large, highly mobile, pelagic carnivores are a real presence in offshore antarctic waters, but the glimpses and insight we have into their biology are, however tantilizing, brief and incomplete. Squid may march to a different beat of the biological drummer.

The decapod cephalopods (squids and cuttlefish) have ten tentacles that include eight prehensile arms equipped with stalked suckers, and two true tentacles, each tipped with a club-like tentaculum. Their shells are reduced or absent in most living forms, although their derivation from shelled ancestors is clearly written in the fossil record, and in the existence of the living pearly nautilis. Squid ancestors include ammonites whose fossil remains are common in the Mesozoic deposits in the Antarctic Peninsula.

There are probably 400 living squid species, although some estimates run as high as 650 species. Squids and cuttlefish of the order Myopsida are denizens of shallow water, continental shelf areas, while squids classed in the order Oegopsida inhabit the pelagic oceanic realm. It is this last group to which most of the squid found in the Southern Ocean belong.

Squid are predaceous, fast swimming animals that feed on a variety of pelagic prey, including copepods and krill, fishes, and other squid. Their

stalked suckers can be armed with horny teeth and hooks that are well adapted to deal with soft-bodied prey. Many oegopsid squid swim at great depths during the day, and rise in schools to the surface at night, where they feed or engage in extravagant courtship behavior.

The large eyes (which are stalked in some of the more abyssal squid) and the ability to change colors rapidly mark squids as highly visual animals. In addition to having nervous system control over their skin color patterns many deep sea squid, including some found in the Southern Ocean, have photophores that can luminesce in ways that could serve to attract food and communicate among peers.

Because of their size, speed, depth, and good eyes, squid are hard to catch, preserve, and maintain in aquariums or museum collections. Lacking bones or other structures that can be used to age them, squid grudgingly give up basic life history information on growth, aging, mortality, and reproductive rates. When feeding they chop their prey into small pieces with their horny beaks and quickly grind the ingested prey with a toothed radula. Digestion seems to be rapid, so that feeding studies based on analyses of stomach contents are difficult to carry out. Under the best conditions most squid are hard to keep alive for laboratory studies. Consequently, our knowledge of the general biology of common temperate species is fragmentary at best. We know even less about the lives of most species of squid in the Southern Ocean.

About 20 species of squid are known from antarctic waters. These range from small, deep-water forms such as *Crystalloteuthis glacialis* to the larger and more surface-oriented *Moroteuthis knipovitchi* and *Kondakovia longimanna*. These latter species, along with the giant squid *Architeuthis* grow to mantle lengths of 3 to 8 meters and often weigh more than 50 to 100 kilograms. Squid this size are ferocious predators, eating krill of the size consumed by baleen whales, as well as amphipods, fishes, and each other.

The larger squid in the Antarctic form a staple for the sperm whale, and smaller sizes are eaten by a variety of birds, including Gentoo and Emperor penguins, seals, albatrosses, and petrels. The overall place of squid in the economy of the Southern Ocean is not known, but it is likely to be important. At least 35 million tons of squid are eaten every year by birds, seals, and whales in the Antarctic. No one knows what fraction of the total squid population this represents, but it does mean that this much squid could account for the demise of more than 100 million tons of krill. Recent studies on many species of squid from the world over reveal that all of them grow rapidly, reaching adulthood in only a year or two, spawning once and then dying. If this kind of life cycle holds true in the Antarctic, squid represent a dramatic contrast to the general pattern of slow growth and great longevity shown by other animals in the far south. In terms of the energy funneled through them, squid may be extraordinarily important. Their proper study is certainly a priority for antarctic biologists.

BENTHIC INVERTEBRATE COMMUNITIES

The fact that the Southern Ocean's fish communities are sparse in species composition and small in biomass might suggest that the benthic invertebrate communities, with which the fish associate, should be so as well. This is not the case. In shallow waters where ice scouring and anchor ice plucking are regular winter-time risks, benthic communities are simple. But below the 33 meter mark (the lower limit of anchor ice formation) the antarctic continental shelf boasts one of the most species-rich, diverse, and dense benthic communities on earth.

The documented richness of this community is probably understated. Extensive near-shore samples have been obtained only in the past thirty years or so, and these are spotty. Like the fishes, many groups of invertebrates here show a high degree of endemism. Lots of species remain undescribed in museum collections; and new species surely are yet to be collected. Still, the Antarctic has yielded at least 875 species of mollusks, 650 species of polychaete worms (57 percent of which are endemic), 470 species of amphipods (90 percent endemism), 299 species of isopods (66 percent are endemic), and more than 310 species of bryozoans. All told the antarctic invertebrate community contains more than twice as many species as the Arctic one.

The benthos is notably rich in species, and it is diverse as well. Year-round sampling at Palmer Station on the Peninsula shows communities that have no single dominating species, and which display species diversities close to the theoretical maximum. This is a reflection of the extraordinary stability of the subsurface marine antarctic environment.

In parts of the Ross Sea and off the western coast of the Peninsula the benthos contains an incredible abundance of sponges, some of which grow to a meter or more in height. Most of these are not the horny, shallow water tropical sponges of commerce, but glass sponges (many in the family Rossellidae) with long, silicious spicules. The spicules of these filter-feeding animals accumulate on the ocean floor to make a mat or felt of glassy particles half a meter thick. They and the living sponges form a porous carpet that gives lots of protection and living space for hosts of other invertebrates like amphipods, pycnogonids, isopods, and mollusks. The substrate enhancement by these sponges and the absence of abundant silts washed off a dry, frozen continent conspire to create a habitat exceedingly rich in number of species, number of individuals, and total biomass. Indeed, the benthic communities found off the Antarctic Peninsula areas rich as any in the world.

Appearances, however, can be deceiving and some caveats are in order. The species abundance, richness, and large populations all signal a thriving invertebrate community that might funnel a large percentage of the Southern Ocean's productivity through to the next trophic levels of predators and decomposers. Such may not be the case. The large biomass of this community is concentrated in sponges, echinoderms, mollusks, and such crustaceans as amphipods, isopods, and mysids. Much of the total biomass of animals like them is made up of minerals in the form of calcium carbonate and other inorganic deposits found in cuticle and chiten. These minerals carry no nutritional value to predators that eat these animals. Certain sea stars, for instance, have ash weights (the inorganic mineral component of body weight) as high as 90 percent of the total; while isopods have ash weights in the neighborhood of 50 percent of the total dry weight. By contrast, ordinary bony fish show ash weights of 6 to 10

''The octopus was a meter across—probably 80 to 90 years old. It would slowly sink to the bottom, spreading its fan of webbed tentacles with the tips curled up. It could quickly change color, from red to deep brownish.''

percent of body weight. What might seem to be abundant food on the bottom of the Southern Ocean turns out to be mainly mouthfuls of chalk!

Many members of this benthic invertebrate fauna share a suite of adaptations that, taken together, show that this is not a land of milk and honey. These adaptations include gigantism, slow growth with great longevity, reduced fecundity, brooding of young, and a startling array of feeding modes that range from filter-feeding planktivory, through active predation, to cannibalism and necrophagy in the same individual.

Biologists who look at collections of antarctic invertebrates frequently remark on the large body size of many endemic species found here. This is particularly true for ''soft-bodied'' animals that use relatively small amounts of calcium, like sponges, crustaceans, polychaete worms, nudibranchs, and ascidians. Thus we find sponges a meter in diameter, a predaceous panthurid isopod *(Accalathura gigantissima)* more than 4.5 centimeters long, 5 centimeter amphipods *(Bovallia gigantae)*, several giant polychaetes, and so on.

In the course of animal evolution gigantism is a recurrent theme—many fish, dinosaurs and other mesozoic reptiles, and a number of pleistocene mammals all tried it. It is usually interpreted as a strategy to grow larger than potential predators, which then are forced to turn to less difficult prey. Once the predators evolve feeding adaptations necessary to handle large prey (as in the case of modern sharks, for example) gigantism ceases to be an advantage, and extinctions occur. Gigantism can be advantageous only if large body size can be achieved quickly through rapid growth. On the surface, therefore, the larger than average size of many Antarctic benthic invertebrates signals a productive, richly growing community.

A closer look, however, shows that this signal is erroneous. These invertebrates are large not so much because their predators are rampant, but because the level of predation here is low. The fish populations, as we have seen, are sparse. Decapod crustaceans, active carnivores of most benthic communities, are not a presence here; nor are benthic feeding birds such as eiders and other sea ducks. Animals whose growth is indeterminent can grow large, even if they grow slowly (as these antarctic forms do), provided they live for long periods of time. Most of the animals in this environment seem to be exceedingly long lived. One starfish *Anasterias rupicola)*, for example, is known to reach ages of at least 39 years.

The heavily calcified invertebrates (echinoderms and bivalve mollusks are good examples) face a real problem of precipitating calcium at the low temperatures of the Southern Ocean. The abundance of calcium is also hindered by the near absence of runoff from the Continent. Accordingly, some of the starfish and other echinoderms have opted for small body size. In contrast to the gigantism found in the soft-bodied and silicious fauna, these animals resort to dwarfism. The starfish, *Granaster nutrix*, which is abundant in shallow water near Anvers Island on the Peninsula weighs less than 1 gram per individual. While most of the echinoderm fauna are larger than this, the larger species of sea stars, crinoids, and sea urchins are generally the ones which have specializations for feeding on krill.

Another recurrent theme in the Antarctic benthos, as in the fish fauna, is a tendency toward reproduction that features small numbers of eggs and extraordinary care of the young. The list of benthic invertebrates that are viviparous is long. It includes many echinoderms, mollusks, ascidians, and even some polychaetes and nemertean worms. Probably the best studied cases of brooding deal with echinoderms.

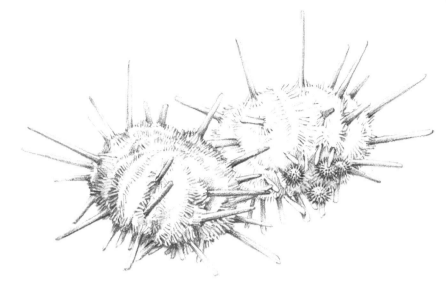

Brooding sea urchin

Some brittlestars, starfish, and other echinoderms from both northern and southern high latitudes retain developing eggs within a maternal body cavity until they hatch and the young develop the adult body form before moving off. In some species (mostly brittlestars) the eggs are nurtured in respiratory bursae. Other echinoderms retain eggs in either the coelom or the ovary. Most Antarctic species seem to retain their embryos for at least 6 to 9 months before the young move away to begin a life on their own. In one brittlestar *(Ophionotus hexactix)* the mother actually transfers nutrients to the developing embryos, either by producing nurse eggs or by body fluids. Brood care in these animals is thus equivalent to the sorts of viviparity found in insects and vertebrates.

The utility of brood protection in these and other benthic dwellers probably has to do with the ephemeral nature of phytoplankton production in the Southern Ocean. Were these animals to cast pelagic larvae into the water column, their slow development would take them into the winter season when primary productivity is virtually nil in the surface waters. It is better to keep a few eggs at home and give them care, than to reproduce in the profligate manner of their temperate and tropical peers. The practice of brooding young also helps the offspring to colonize a favorable habitat. Given the small continental shelf area of the Antarctic, the chances are good that pelagic larvae would be swept to deep water where they would have little chance to settle on a decent bottom.

Stomach contents show that many Antarctic marine invertebrates feed opportunistically. Many brittle stars, crinoids, mollusks, and sea urchins capture and eat krill when they are available. During the dark of the year when krill move offshore into deeper water these same invertebrates switch

their dietary preferences to scavenge just about anything available. One little starfish, *Anasterias rupicola*, even engages in cooperative feeding behavior where several individuals will gang up on one much larger prey, a limpet.

Given the high productivity of the pelagic world of the Southern Ocean

Limpet

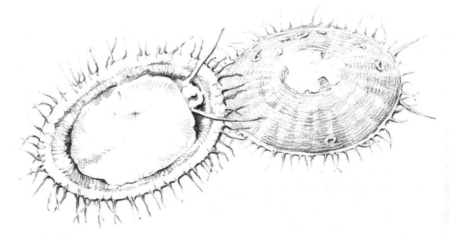

with its abundance of diatoms, and the krill and copepods that feed on it, one properly expects the benthos to show a proportionate luxuriance. Its abundance and richness of species conforms to this expectation. But life here is slow. The crumbs that fall to the bottom nurture an existence that is meager and eked out only by dint of special adaptations that are recurrent in many of the animals living here. It is life in slow motion.

To look at a faster paced life, and one that features adaptations of different sorts, we go now to the surface of the Southern Ocean, and the birds there that rely primarily on krill for their sustenance.

Amphipod

ADDITIONAL READING

Ahlgren, J. A. and A. L. DeVries. 1984. Comparison of antifreeze glycopeptides from several antarctic fishes. *Polar Biology,* 3:93–97.

Akimushkin, I. I. 1963. *Cephalopods of the Seas of the U.S.S.R.* Translation. Jerusalem, 1965.

Andriashev, A. P. 1965. A general review of the Antarctic fish fauna. In J. van Mieghem and P. van Oye, eds., *Biogeography and Ecology in Antarctica,* pp. 303–402. The Hague: Junk.

Arnaud, P. M. 1977. Adaptations within the Antarctic marine benthic ecosystem. In G. A. Llano, ed., *Adaptations Within Antarctic Ecosystems,* pp. 135–157. Washington, D.C.: Smithsonian Institution.

Blankly, W. O. and G. M. Branch. 1984. Co-operative prey capture and unusual brooding habits of *Anasterias rupicola* (Verrill) (Asteroidea) at sub-Antarctic Marion Island. *Marine Ecology Progress Series,* 20:171–176.

Boyle, P. R., ed. 1983. *Cephalopod Life Cycles.* Vol. 1, *Species Accounts.* Orlando, Fl.: Academic Press.

Briggs, J. C. 1974. *Marine Zoogeography.* New York: McGraw-Hill.

Burchett, M. S. 1983. Abundance of the nearshore fish population at South Georgia (Antarctica) sampled by trammel net. *British Antarctic Survey Bulletin,* 61:39–43.

Clarke, A. 1984. Lipid composition of two species of *Serolis* (Crustacea, Isopoda) from Antarctica. *British Antarctic Survey Bulletin,* 64:37–53.

Clarke, A., N. Doherty, A. L. DeVries, and T. T. Eastman. 1984. Lipid content and composition of three species of Antarctic fish in relation to buoyancy. *Polar Biology,* 3:77–83.

Clarke, M. R. 1983. Cephalopod Biomass—Estimation from predation. *Memoirs of the National Museum, Victoria* 44:95–107.

Cohen, D. M. 1970. How many recent fishes are there? *Proceedings of the California Academy of Sciences, Series 4,* 38(17):341–346.

Daniels, R. A. 1982. Feeding ecology of some fishes of the antarctic peninsula. *Fishery Bulletin,* 80:575–588.

Dell, R. K. 1972. Antarctic benthos. *Advances in Marine Biology,* 10:1–216.

DeVries, A. L. and Y. Lin. 1977. The role of glycoprotein antifreezes in the survival of Antarctic fishes. In G. A. Llano, ed., *Adaptations Within Antarctic Ecosystems,* pp. 439–459. Washington, D.C.: Smithsonian Institution.

DeWitt, H . H. 1968. Coastal fishes. *Australian Natural History.* Dec., pp. 119–123.

DeWitt, H. H. and T. L. Hopkins 1977. Aspects of the diet of the Antarctic silverfish, *Pleurogramma antarcticum*. In G. A. Llano, ed., *Adaptations Within Antarctic Ecosystems*, pp. 557–567. Washington, D.C.: Smithsonian Institution.

Douglas, E. L., K. S. Peterson, J. R. Gysi, and D. J. Chapman. 1985. Myoglobin in the heart tissue of fishes lacking hemoglobin. *Comparative Biochemistry and Physiology, 81A(4):885–888.*

Eastman, T. T. and A. L. DeVries. 1982. Buoyancy studies of notothenioid fishes in McMurdo Sound, Antarctica. *Copeia*, no. 2, pp. 385–393.

Everson, I. 1984. Fish biology. In R. M. Laws, ed., *Antarctic Ecology*, 2:491–532. London and Orlando, Fl.: Academic Press.

Fratt, D. B. and J. H. Dearborn. 1984. Feeding biology of the antarctic brittlestar *Ophionotus victoriae* (Echinodermata: Ophiuroidea). *Polar Biology*, 3:127–139.

Hemmingsen, E. A. and E. L. Douglas. 1970. Respiratory characteristics of the hemoglobin-free fish *Chaenocephalus aceratus. Comparative Biochemistry and Physiology*, 33:733–744.

Hemmingsen, E. A. and E. L. Douglas. 1977. Respiratory and circulatory adaptations to the absence of hemoglobin in chaenichthyid fishes. In G. A. Llano, ed., *Adaptations Within Antarctic Ecosystems*, pp. 470–487. Washington, D.C.: Smithsonian Institution.

Hubold, G. 1984. Spatial distribution of *Pleurogramma antarcticum* (Pisces: Nototheniidae) near the Filchner—and Larsen Ice shelves (Weddell Sea/Antarctica). *Polar Biology*, 3:231–236.

Hureau, J. C., D. Petit, J. M. Fine, and M. Marneux. 1977. New cytological, biochemical, and physiological data on the colorless blood of the Channichthyidae (Pisces, Teleosteans, Perciformes). In G. A. Llano, ed., *Adaptations Within Antarctic Ecosystems*, pp. 459–478. Washington, D.C.: Smithsonian Institution.

Johnston, I. A. and J. D. Altringham. 1985. Evolutionary adaptation of muscle power output to environmental temperature: Force-velocity characteristics of skinned fibres isolated from antarctic, temperate and tropical marine fish. Pflugers Archiv, 405:136–140.

Kock, K.-H., R. Schneppenhesim, and V. Siegel. 1984. A contribution to the fish fauna of the Weddell Sea. *Archivs Fischwissenschaft*, 34(2/3):103–120.

Laws, R. M. 1985. The ecology of the Southern Ocean. *American Scientist*, 73:26–40.

Montgomery, J. C. and J. A. Macdonald. 1984. Performance of motor systems in Antarctic fishes. *Journal of Comparative Physiology*, A 154:241–248.

Targett, T. E. 1981. Trophic ecology and the structure of coastal antarctic fish communities. *Marine Ecology Progress Series*, 4:243–263.

Turner, R. L. and J. H. Dearborn. 1979. Organic and inorganic composition of post-metamorphic growth stages of *Ophionotus hexactis* (E. A. Smith) (Echinodermata: Ophiuroidea) during intraovarian incubation. *Journal of Experimental Marine Biology and Ecology*, 36:41–51.

Umminger, B. L. 1977. Mechanisms of cold adaptation in polar marine animals. In G. A. Llano, ed., *Adaptations Within Antarctic Ecosystems*, pp. 397–408. Washington, D.C.: Smithsonian Institution.

Wagele, J. W. 1985. Observations on nutrition and ultrastructure of digestive tract and fat body of the giant paranthurid *Accalathura gigantissima* Kussakin. *Polar Biology*, 4:33–43.

Ward, P. 1984. Aspects of the biology of *Antarctomysis maxima* (Crustacea: Mysidacea). *Polar Biology*, 3:85–92.

White, M. G. 1984. Marine benthos. In R. M. Laws, ed., *Antarctic Ecology*, 2:421–461. London and Orlando, Fl.: Academic Press.

CHAPTER 6

SECONDARY CONSUMERS— BIRDS

In addition to predation by the fish, squids, and benthic invertebrates of the sea, great numbers of euphausiids and other herbivorous plankton are eaten by birds that live in and above the Southern Ocean. Indeed, the biomass of birds south of the Antarctic Convergence today probably exceeds 201 milion metric tons. Altogether these birds eat 14.6 trillion kilocalories of energy per year. About 78 percent of these calories are gotten from krill, with most of the remainder coming from fish and squid.

Although the Southern Ocean sets its table for as many as 67 species of birds, they, like the fishes that swim below them, are drawn from only a few orders. Prominent among these are the penguins (Sphenisciformes) and the petrels (Procellariiformes), with a few gulls and terns (Charadriiformes) and shags (Pelecaniformes) also present. Most species breed on the more equable islands of the Subantarctic, at or not far south of the Antarctic Convergence (South Georgia, Bouvet, Crozet, Kerguelen, Marion Isles, etc.). Relatively few of them are adapted for the rigors of nesting and rearing young in the harsher conditions presented by the Antarctic

Peninsula and East Antarctica. It is on these more hardy species that our attention will be focused. Antarctic breeding birds that are primary carnivores include five petrels, four penguins, and a tern. In addition to these, several other species make most of their living preying on, or scavenging the remains of other predators. These secondary carnivores will be discussed in chapter 8.

PENGUINS

One group of birds that most people associate with the Antarctic are the penguins (Order Sphenisciformes, family Spheniscidae). Although the group is small—eighteen species are recognized—and confined to the Southern Hemisphere, only five species are truly Antarctic in the sense that they breed on the Continent, adjacent islands, or on fast ice anchored to it. The remaining penguins breed on islands not far from the Antarctic Convergence as well as along the coasts of South Africa, southern Australia, New Zealand, Patagonia, Chile, and Peru as far north as the Galapagos Islands. Excepting the King penguin of subantarctic islands and the Yellow-eyed penguin of New Zealand, the truly Antarctic penguins are the largest species in the order, and show special physiological, behavioral, and morphological adaptations to deal with the extreme environment in which they thrive.

Well-defined penguin fossils (usually consisting of isolated wing and leg bones) are known from the major southern Continents—including the Peninsula—dating back to the Eocene epoch, some 40 million years ago. Most authorities agree they stem from flying ancestors, not the ratite line that produced modern ostriches and emus. Biochemical evidence suggests

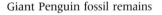

Giant Penguin fossil remains

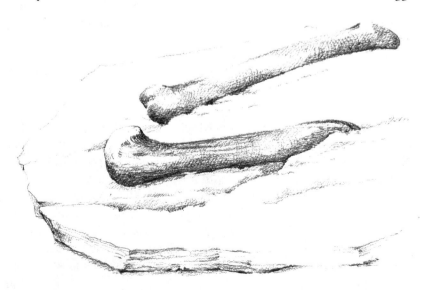

that they share a common ancestry as well as the Southern Ocean with the petrels. Other hypotheses give them an ancestry from auk-like birds. Whatever their evolutionary affinities the penguins—and especially the Antarctic ones—are unusual, spectacular birds.

Adélie Penguin

The most abundant (and best studied) of the Antarctic species is the Adélie penguin *(Pygoscelis adélie)* the nesting colonies of which are common and accessible in many places along the coastline of the Peninsula. The life history of these birds begins with the formation of breeding rookeries often composed of tens and sometimes hundreds of thousands of breeding pairs. Adélies, which range from the more southern islands of the Scotia Arc to the entire coastline of the antarctic continent, choose windswept, exposed ground for their breeding colonies, probably because these are the locations most apt to be cleared of snow early in the austral summer. The penguins troop to traditional rookery sites well before the inshore fast ice has broken up. This for most localities on the Peninsula occurs in late October, and causes the birds to march across significant distances of ice to reach the shoreside rookeries. In East Antarctica breeding sites exist that require an over-ice and land journey in excess of 100 kilometers, but 20 to 40 kilometers is more like the norm. These are impressive treks for birds that stand about 70 centimeters high and weigh 4.5 to 5.5 kilograms on the average.

In moving across fast ice and land the Adélies travel in single file with their curious rolling gait at a rate of about 8 kilometers per hour. If the snow, ice, and grade are favorable, or when the birds are harried, they

"What funny birds! Many are on shore, climbing around the ice, others hopping out of the big surf smashing against the rocks. Somehow they pop out of the water and land on the rocks. Then they shake like dogs and preen themselves. They seem so clumsy on their feet— falling on their bellies, hopping, waddling, and tripping. But they keep charging ahead in their clumsy way and cover a lot of ground quickly when they want to. Even on flat ground or uphill they sometimes dive onto their bellies and kick with their feet."

"Many Adélies are courting, raising their heads and beaks, swaying from side to side facing each other, and cackling."

can toboggan along on their bellies, with their pink, black-soled feet propelling them along at an even faster clip. When the penguins really want to make speed, they bring their flippers into play, and can toboggan faster in the snow than a man can run under the same conditions.

Tagging studies show that individual Adélies not only return to the same rookeries of their birth, but occupy the same nest sites year after year. When the penguins reach these sites the males stake out territories on whatever bare ground exists, and for the next three weeks engage in courtship, mating, territorial disputations, and gathering pebbles and small stones with which to build their nondescript nests. At this time the birds are at the peak of their physical vigor and appearance.

Adult Adélies are easy to identify with their blue-black heads and chins, pure white underbellies and distinctive white-lidded eyes. The male and female are superficially similar, although males average heavier in weight and have larger bills and mouths. Standing erect on their fleshy feet, the long stiff tails characteristic of the whole genus prop them up tripod-like, allowing them to doze off and sleep while standing erect. Even more than their congenors, the Chinstrap and Gentoo penguins, the Adélies are adapted for the exceptionally harsh conditions in which they are found. The thick pelt of extraordinarily dense feathers is underlain by a substantial layer of subcutaneous fat that serves both as insulation and a hedge against incubation, courtship, and molting periods when the birds go weeks without eating. Even the bill is uniquely feathered for half its length, and the external nares are tightly closed to prevent the unwanted loss of body heat.

Feathered beak of Adélie Penguin

In their handsomely marked plumage and droll, swaying gait, these penguins never fail to impress the human onlooker as anthropomorphic improvisations of fussy little men (even though fully half of them are females). Behaviors including courting and territorial displays, agonistic interactions, and disputes over the ownership of nesting pebbles often seem comical to the casual observer. For the penguins, however, this is serious stuff. The central purpose is reproduction, and when the environment is as hostile as this one, successfully raising chicks to fledging is not an easy undertaking.

While penguins are attractive, active, and appealing birds, the reality of a hundred thousand of them at once can be overwhelming. On one hand there is the behavioral biology to enjoy—the displays, the fights, the furtive kleptomania of pebble thievery—but on the other is an incredible din of one hundred thousand groans, squawks, shrieks, and peeps. Some of the exceptionally large Adélie rookeries in the McMurdo Sound region can be heard as far away as 50 kilometers downwind. Then there is the odor. The fecal production of a half a million kilograms of bird biomass is prodigious. Add this stench to that of large amounts of spilled regurgitated chick feed and it is understandable why one, on entering a penguin colony, encounters a wall of ammoniacal pungence that drains from the sinuses for days afterward!

After a three-week period of courtship, mating, and nest building, the females settle down to lay their eggs (averaging two per nest) and incubation begins. The male usually takes the first duty (some females may take a short first shift), which lasts about two weeks, while the female returns to the water to feed and recover the weight lost during courtship and egg laying. She eventually comes back to the nest to relieve the male, and after a short period of mutual displaying, he departs for his own fiftcen-day stint of gluttony.

''Their eyes remind me of those little plastic eyes on stuffed animals with the moving black disk inside. It makes them look comical and cartoonish. But when they display to intruders they roll the eye down, exposing more white to make it even more prominent.

Adélie eggs in nest

Incubating Adélie

Adélie penguins incubate their eggs as do all but King and Emperor penguins by lying belly down on them, warming them with a well-vascularized and fat-rich section of the belly skin—the brood patch. While the normal body temperature of the adults is within a degree or two of 39°C, the egg temperature during incubation is closer to 30°C. This is lower than the incubation temperatures for many birds and reflects the poor insulating qualities of stones as a nesting material. The eggs weigh a little less than 120 grams on the average when they are laid. From each a 90 gram chick emerges after an incubation period of about 36 days.

As hatching time nears the female returns to the nest and is usually on the eggs when they hatch. If the eggs are laid in the first half of November (normal for most Peninsula breeding sites), the eggs begin hatching in the third week of December. Synchrony in egg laying is normal in Adélies, but some fertile eggs may be found as late as the middle of January. After the eggs hatch the male and female trade shifts brooding the young on a much shorter schedule, each turn averaging between two and three days. The gray, downy chicks grow rapidly on a diet of krill regurgitated for them by the foraging parent, and daily weight gains average 100 or more grams throughout the 50-day period before fledging.

After leaving the nest the penguin fledglings cluster together in groups known as creches, where each animal gains protection from the group against predators such as skuas and Southern Giant Petrels. During periods of particularly inclement weather the creche also provides a buffer to the hostile environment. During this period both parents forage at sea, returning occasionally to provision their own chick which they somehow locate within the creche structure.

When the parent identifies its chick, it turns and runs away, forcing the hungry youngster to follow. Alien chicks also join in the chase, but usually only the correct offspring is persistent enough to be finally rewarded by the doting parent.

The character of a good-sized penguin colony changes as the chicks develop. Fecal matter accumulates from both old and young birds. Piles

Newborn Adélie chick

*"The young are funny—flippers dispro-
portionately long, belly huge, all rag-
ged, brown and fluffy—clumsy floppy
things. This big, tangled lump of helpless bird lying at the feet of its parent
that is all crisp and groomed and alert
by contrast. The chicks are nothing but
digestive tubes—feeding machines. I
suppose their mothers think they're cute!"*

Adélie creche

of reddish krill, prematurely regurgitated and then uneaten, litter the ground. The adults become emaciated as the young grow, and their once pristine plumage becomes stained with offal. Mortality in chicks is high, particularly in those being reared near the periphery of the colony by younger, inexperienced parents.. Predaceous and scavenging birds, the

"The young now have half of their fluffy down gone. The juvenile plumage underneath is like the adult's, but with a white chin. It's a sleek coat compared to the down. They look like clipped French poodles with blobs of fluff here and there. Often the down on top of the head stays a while—looks like a wig."

skuas, giant petrels, and sheathbills crowd in, harrying the chicks, stealing eggs, and feasting on the carcasses of young abandoned or otherwise unprotected by their parents. What began as a startling and somewhat comical procession of neat, plump, and garrulous birds gives way to a soiled, sodden battlefield, where the drive for genetic survival becomes stark indeed.

The brownish first set of feathers acquired by the chicks clearly identifies them as juveniles. Having been abandoned by their parents they finally take to the water and begin to forage on their own. Adélie penguins in their first year of life are rarely seen near the Peninsula, and for this reason their movements are poorly known. The young penguins probably range further north in the Southern Ocean than do the adults. When females are in their third year of life and males in their fourth, they begin the return to their natal rookeries to attempt to breed. The average age of maturity, however, is closer to 5 years for females and 7 years for males. Like most birds in the Antarctic, penguins are long-lived. Adélie penguins, however, suffer higher annual mortality than most other penguins (between 30 and 40 percent of breeding-age adults succumb every year). Their large population size reflects a steady, rich food supply, and a potent reproductive biology.

After the chicks are safely fledged, the worn and weary adults depart for the offshore feeding grounds to recover the weight which they committed to chick rearing, and to prepare them for the metabolic rigors of molting—the final summertime hurdle that they must cross. In a typical two-week turn at incubating eggs an Adélie penguin loses between 600 and 900 grams of weight, or about 20 percent of its initial body weight. After regaining this lost weight during a three-week sojourn to the feeding grounds, the adults return to the colony to sit quietly while their old plumage is shed and a new set of feathers grows in. This consumes a lot of energy, for while the birds' physical activity is minimal at this time, the production of feathers requires a great deal of protein synthesis.

Because the penguins cannot go to sea during this process the energy for the molt must come from fat and other substances stored in the body. Molting causes yet another period of weight loss; this time more extreme than that experienced during incubation and brooding. In a twenty-day molt period, for example, the typical Adélie penguin loses more than 150 grams of weight per day, or over 3 kilograms during the molt. This is close to half its initial body weight. Fortunately for the emaciated Adélie emerging from the far side of its molt, the early fall waters off the Peninsula teem with krill, both *Euphausia superba* and *E. crystallorophias*, and the penguins are able to prepare their bodies for winter after only a short feeding period.

It is in the water, rather than on land, that all penguins shine. Their

Molting Adélie adult

sleek, dense feathers streamline them for maximum hydrodynamic efficiency. The fat, pudgy-seeming penguin shape actually fits the ideal 4.5 fitness ratio (length/width) that is a prerequisite for drag reduction; and the hydrofoil-shaped flippers (wings) are driven by an extraordinarily powerful musculature. As a result penguins literally dart about under water, and the Adélies streak through swarms of krill, zigzagging as they go, plucking the rich crustaceans. Flocks of penguins normally coordinate their

underwater movements, maintaining visual communication with each other—a process that is enhanced by the contrasting white and black markings on the body. Although they can remain submerged for as long as six minutes, the birds normally return to the surface each minute or two in order to breathe. When on the move as in feeding or migrating the penguins breach in "porpoising" fashion, throwing themselves clear of the water as so many rapidly moving missiles. During periods of lesser activity the birds may loll at the surface with their heads held above the waves.

Perhaps the most astonishing aspect of their swimming behavior is the manner in which they come ashore when faced with an abrupt, cliff-like strand. The penguins seem to gauge the distance and elevation of the shoreline from a distance of perhaps 30 meters seawards. Then swimming rapidly under water, they rush to land's edge and "bounce" straight up out of the water to land on their feet atop sheer drops of ice or rock as high as 2 meters above the water's surface. They have an uncanny ability to know whether to land on their feet (if the surface be rock or rough ice) or in tobogganing attitude on their bellies (if the surface is smooth ice or snow).

Gentoo Penguin

While Adélie penguins are the most abundant penguins of the true Antarctic, two closely related species, the Gentoo and Chinstrap penguins, are frequently found near them along the northern half of the Peninsula. Of these two the Gentoo *(Pygoscelis papua)* is the most dissimilar. Gentoos are larger (nearly 6 kilograms in average body weight) than Adélies, and are conspicuously marked with white patches on either side of the top of the head. While they often form nesting colonies in the same rookeries with Adélies, they generally choose flatter ground, nearer the water. They

Sheathbill in Gentoo nesting colony

Gentoo

build larger nests that are more widely spaced (more than a meter apart) than are those of the Adélies. Because Gentoo breeding colonies are more spread out, and adults seem less alert than Adélies, scavenging birds like American Sheathbills and Southern Giant Petrels are more likely to walk right through the center of them, posing definite threats to the penguin eggs and chicks. The more tightly packed Adélie colonies discourage such intrusions.

In their feeding biology Gentoos show differences from the other *Pygoscelis* species. While they eat mainly krill, a minor component (as much as 15 percent) in their diet off the Peninsula is fish and squid. Where Adélie and Chinstrap chicks fledge in only 50 to 54 days after hatching, Gentoos require 70 to 90 days to fledge. Add this to the 36 days of incubation and a couple of prior weeks of courtship and nest site selection, and the total reproductive season stretches out to 120 to 145 days. One payoff from this is that the chicks are larger at fledging than are those of the Adélies and Chinstraps. But this is perilously long for the Antarctic, and could help to explain why the Gentoo breeds not much further south than about 65° along the Peninsula.

Chinstrap Penguin

The third *Pygoscelis* species of penguin is the Chinstrap, *P. antarctica.* Smaller than either of its close relatives (body weight averages a little over 4 kilograms) and named for the distinctive narrow black band of feathers that passes under the white chin, this penguin shows real climbing ability. In those rookeries on the Peninsula where all three of these penguins form mixed nesting colonies, the Chinstraps choose the highest, most rocky sites for their nests. The birds use their beaks and claws to pull and push

"The chinstraps greeted each other in a manner similar to the Adélies, but with a slightly different voice. After a couple of minutes he reached over toward the two eggs, and they switched places with much chatter."

their way to rock pinnacles as high as hundreds of meters above sea level in some areas. The nests are rudimentary in such places, consisting of just enough pebbles and small rocks to keep the two eggs from rolling off the precipice.

The Chinstrap penguin has a reproductive biology not much different from the Adélie. But it breeds in greatest abundance in the northern reaches of the Peninsula and through the southern islands of the Scotia Arc. Of these three species of penguins, we know the most about the foraging behavior of Chinstraps. They feed mainly on krill which they capture on dives that have been recorded as deep as 70 meters. Actually, most of their dives are to much lesser depths, 90 percent being to 45 meters or less. A single dive off Signy Island in the South Orkneys where these measurements were made, lasts about 1.6 minutes, and produces for the penguin about 16 krill on the average. The birds make between seven and fourteen dives per hour, ultimately bringing back more than a pound of krill with which to feed their young.

Although these three species of penguins are by far and away the most common small penguins breeding on the Peninsula, other penguins have been reported here. Wandering non-breeding Rockhopper penguins *(Eudyptes chrysocome)* and Macaroni penguins *(E. chrysolophus)* have been recorded at Palmer station. These are crested penguins, small in body size,

"The Macaroni has some striking markings—particularly the yellow crest feathers, the red eye, and a thick rusty-orange beak, pink in the corner."

but which carry distinctive yellow plumes above their eyes. In the summer of 1985–86 a pair of Macaroni penguins attempted to breed near Anvers Island. Their presence in the South Shetlands is well established, so their breeding range may be extending south.

The remaining antarctic penguin, the spectacular Emperor penguin *(Aptenodytes forsteri)* is more likely to eat squid and fish than krill. It, therefore, is a secondary carnivore and will be considered in chapter 8 with the other birds of that trophic level.

PETRELS AND TERNS

Despite their great numbers and large collective biomass, penguins are not the only krill-eating birds that breed on the Antarctic Peninsula. Several species of petrels and a tern or two are also conspicuous in their presence here and interesting in their natural history.

Petrel's tube nose

Various species of albatrosses (Diomedeidae) and petrels (Procellariidae) are the most numerous and ubiquitous birds of the Southern Ocean. These graceful and fascinating creatures range in size from the huge albatrosses, unequaled in their gliding aerial artistry, to the tiny fluttering storm petrels that weigh but 40 grams (less than two ounces) as breeding adults. As a group of birds petrels share the characteristic "tube" nose—a gun-barrel-shaped elevation atop the bill that exaggerates the external nares. The ease with which many species of petrels find the greasy sites of seal and whale butchery points up the fact that they, unlike many other birds, have a keen sense of smell.

Another adaptation, common to non-diving petrels, is the ability to store in the stomach a quantity of rich, foul-smelling oil that ranges in color in different individuals from bright red to bilious green. Both adults and chicks in the nest are able to forcefully eject this oil a considerable distance—two meters or more—much to the dismay of scientists visiting breeding colonies of these birds. The origin and function of this oil has long been debated among ornithologists. Its chemical composition suggests a direct derivation from the food that the petrels eat. Histological studies of the stomach wall, however, suggest that it is secreted into the stomach by the cells which line that organ. What is clear is that the birds feed it to their nestlings, and it provides a concentrated high-energy source of food that is more convenient to carry long distances from foraging grounds to nests than would be the actual prey eaten by the petrels. Its second clear function is as a defensive armament. Predatory birds like skuas, once drenched with this oil from a defensive brooding petrel or its chick, can lose the insulating integrity of its own plumage and quickly succumb to the cold temperatures of the antarctic environment.

Despite the numerical abundance of petrels in the Southern Ocean, relatively few species are able to breed successfully in the far southern latitudes of the Peninsula and East Antarctica. Of these the ones that rely on krill or other zooplankton as a primary source of food can be counted on the fingers of one hand. They are: the Cape Petrel *(Daption capense)*, Southern Fulmar *(Fulmarus glacialoides)*, Snow Petrel *(Pagodroma nivea)*, Wilson's Storm Petrel *(Oceanites oceanicus)*, and, possibly, the Antarctic Petrel *(Thalassoica antarctica)*. A sixth species, the Dove Prion *(Pachyptila desolata)*, was recorded breeding in East Antarctic only once, and that in 1912. Other birds that could be expected to breed in peninsular waters include the Black-bellied Storm Petrel *(Fregetta tropica)* and the Blue Petrel *(Halobaena caerulea)*.

All of these petrels feed principally (although not exclusively) on krill and other crustaceans gathered in the surface layers of water, usually not far from the pack ice. Indeed, high relative densities of some of them at sea are good indicators of the presence of shoals of krill. All are opportunistic feeders in that they will eagerly partake of carrion such as pieces of blubber or bird and fish carcasses when they are available. A final unifying characteristic is that while these petrels are adept masters of the windy, turbulent air over the Southern Ocean, and swim like corks upon its surface, they are well-nigh helpless when on land. All of these birds can scarcely stand erect when ashore at their nests, and walk with only the greatest difficulty by dragging themselves forward with their wings in kind of a "quadrapedal shuffle."

The importance of zooplankton in their diets can be appreciated most clearly by looking at the feeding apparatus of such birds as the Cape Petrel and Dove Prion. They have upper and lower mandibles that are lined

"Cape petrels glide into the wind, motionless almost to a standstill, then bank slightly to one side and pick up speed as they fall off with the wind; their backs turning into the sun showing the unmistakable black and white blotches on their backs and wings."

"I saw Wilson's Storm Petrels about eleven o'clock when the wind was fierce. They were circling and zigzagging like swallows, and sometimes slipping in between the rocks to their nests. It snowed two inches."

on their lateral edges with serrations that serve to strain krill and other plankton from the water. In their feeding they spread their wings and paddle forward with their feet over the surface of the water like little hydroplanes, scooping mouthfuls of water that are then filtered. Occasionally the birds make brief dives beneath the surface. The entrapped krill gather in a pouchlike sac in the lower jaw until enough have accumulated to make swallowing them worthwhile.

With one exception these petrels are birds of the Southern Hemisphere. In the austral winter the Cape Petrel and Southern Fulmar are known to move northward, sometimes approaching the equator off the west coast of South America. The real voyager is Wilson's Storm Petrel. This little dark bird with a conspicuous white rump patch moves well into the Northern Hemisphere, at least to 50° latitude in the Atlantic, where it is well-known to fishermen as "Mother Carey's Chicken." The small size of this bird and its fluttering flight make such an annual migration seem truly remarkable.

Many interesting aspects of the biology of these southernmost elements of the world's avifauna have to do with their reproductive natural histories. The emphasis is on the plural here for this handful of petrels conform to two quite different nesting and brood-rearing strategies—both of which turn out to be successful in avoiding the twin dangers of predation and the hostile weather.

The largest of these petrels, the Cape Petrel (body length, 40 centimeters; wing span, 80 to 90 centimeters; total weight about 250 to 300 grams), the Southern Fulmar (40 centimeters long; wingspan, 150 centimeters; body weight, 490 grams), and the Antarctic Petrel tend to nest in colonies that sometimes number in the millions of birds. The mutual stimulation resulting from this behavior provides synchrony in courting, egg laying, and hatching. As in other synchronous animals this serves to present populations of predators with a food supply (the eggs and chicks) that exceed their immediate needs. In this way each set of parents stands a reduced chance of losing its egg or chick (usually only one egg is laid) to predators.

Along with synchronous breeding behavior is the habit of nesting earlier in the summer than do the smaller species of petrels. The nesting sites of these birds, which tend to be ledges on inaccessible cliffs, are likely to be swept clear of snow before the more sheltered sites chosen by the prions and storm petrels. Having laid their eggs, the petrels incubate them in shifts averaging three or four days. While one parent sits on the egg, the other forages to maintain its body weight and to prepare for the appearance of a ravenous new chick. After an incubation period of about 45 or 46 days the egg hatches and the young is fed on stomach oil and krill until it fledges about a month later. After a few days of foraging the adults enter their molt and then get ready for the fast approaching winter. Of these colony nesters the Antarctic Petrel is perhaps unique, for it alone chooses nesting sites well inland (often more than 100 kilometers from the coast) on the East Antarctic Continent.

The smaller species of petrels (Snow Petrel, Dove Prion, and Wilson's Storm Petrel) use a somewhat different strategy to deal with the Antarctic's uncertainties. These birds seek shelter for their nests. In the islands of the Scotia Arc where soils are better developed, Wilson's Storm Petrel and the Dove Prion excavate burrows as much as a meter into the soil. On the Peninsula, however, such opportunities are rare and natural crevices in rocks or talus slopes are sought for nesting sites. One result is that the nests are scattered and breeding synchrony is foregone.

With snow being a major cause of chick mortality, a protracted nesting season allows either early or late nesters to escape unusual snowfalls in a given year. Fledging in late hatching Wilson's Storm Petrels may not occur until early April, just a week or two before the adults depart for their northward migrations.

Crevice holding Storm Petrel's nest

Crevice and burrow nesting petrels have evolved a definite nocturnal habit. Visits to the nests are made only at night when predators such as skuas and giant petrels are less active. This in turn has made the location of these nests difficult for ornithologists who study them. Although the problem has not occurred on the Peninsula or East Antarctica, many populations of burrowing petrels have been decimated on some of the subantarctic islands, such as Crozet and South Georgia, because of unwitting animal introductions by man. The culprits are Norway rats and domestic cats that jumped ship from bygone whalers and sealers. The rats and cats found conditions to their liking here, particularly the food source afforded by burrowing petrels. Consequently hundreds of thousands, if not millions, of these birds are slaughtered annually by these nocturnally active predators. It is to be hoped that this kind of exotic introduction will not succeed on the Antarctic continent proper.

Storm Petrel chick

The single krill-feeding, non-petrel that breeds on the Antarctic Peninsula is the Antarctic Tern, *Sterna vittata*. This is a nonmigratory resident that feeds on the cryopelagic community at the edge of the pack ice in winter and in open water on Antarctic silverfish in summer. Its appearance is similar to the Arctic Tern that breeds in the far Northern Hemisphere and which migrates to the Antarctic for the austral summer. The easy interpretation would be that it's just an Arctic Tern that decided not to migrate 12,000 or 13,000 kilometers north. Its juvenile plumage, however, is distinctly different from its northern, migratory relative, so its distinctiveness is due to more than mere migratory sloth.

Antarctic Tern

The Antarctic Tern is an early nester that lays in the first half of November at Anvers Island on the Peninsula. Because it nests in exposed

"Nest" and eggs of Antarctic Tern

locations, predation on eggs and chicks in not unusual, although by breeding so early it beats the main predatory pressure that occurs when skuas have their own nestlings to feed. This bird, unlike its petrel peers, will relocate and lay another clutch should its first prove unsuccessful.

Newly hatched Antarctic Tern chick

"The tern chicks are in many stages. Lots of eggs are still unhatched, there are day-old chicks, and more developed ones with most of their feathers in. Some have already left the nest. Each stage of growth offers different textures and patterns on the feathers."

This means that the population can have a very long breeding season, and, indeed, eggs can still be found as late as the beginning of February.

Even though the flying birds of the Antarctic region are numerous and conspicuous, particularly in the summer, their total biomass is but a small fraction of that estimated for the stocks of penguins that live in the Southern Ocean. Recent calculations of seabird populations in the Antarctic region show a total biomass of 201 million metric tons, only 3 metric tons of which are birds other than penguins. These flying sea birds, mainly petrels, consume only 0.5 trillion kilocalories of energy (mainly as krill) per year, while the penguins account for 14.1 trillion kilocalories. It will be worthwhile to examine, in the next chapter, the portion of this resource that is utilized by the mammals, the seals and whales, of the Southern Ocean.

ADDITIONAL READING

Ainley, D. G. and D. P. DeMaster. 1980. Survival and mortality in a population of penguins. *Ecology,* 61:522–530.

Ainley, D. G. and W. B. Emison. 1972. Sexual size dimorphism in Adélie penguins. *Ibis,* 114:267–271.

Ainley, D. G., R. E. Le Resche, and W. J. L. Sladen. 1983. *Breeding Biology of the Adélie Penguin.* Berkeley: University of California Press.

Beck, J. R. 1970. Breeding seasons and molt in some smaller Antarctic petrels. In M. W. Holdgate, ed., *Antarctic Ecology,* 1:542–550. New York: Academic Press.

Carrick, R. and S. E. Ingham. 1970. Ecology and population dynamics of Antarctic sea birds. In M. W. Holdgate, ed., *Antarctic Ecology,* 1:505–525. New York: Academic Press.

Clarke, A. and P. A. Prince. 1976. The origin of stomach oil in marine birds: Analyses of the stomach oil from six species of subantarctic procellariiform birds. *Journal of Experimental Marine Biology and Ecology,* 23:15–30.

Croxall, J. P. 1982. Energy costs of incubation and molt in petrels and penguins. *Journal of Animal Ecology,* 51:177–194.

Croxall, J. P. 1984. Sea birds. In R. M. Laws, ed., *Antarctic Ecology,* 2:533–616. London and Orlando, Fl.: Academic Press.

Kooyman, G. L., R. L. Gentry, W. P. Bergman, and H. T. Hammel. 1976. Heat loss in penguins during immersion and compression. *Comparative Biochemistry and Physiology,* 54A:75–80.

Laws, R. M. 1977. The significance of vertebrates in the Antarctic Marine Ecosystem. In G. A. Llano, ed., *Adaptations Within Antarctic Ecosystems,* pp. 411–435. Washington D.C.: Smithsonian Institution.

Laws, R. M. 1985. The ecology of the Southern Ocean. *American Scientist,* 73:26–40.

Lishman, G. S. and J. P. Croxall. 1983. Diving depths of the Chinstrap penguin *Pygoscelis antarctica. British Antarctic Survey Bulletin,* 61:21–25.

Obst, B. S. 1985. Densities of antarctic sea birds at sea and the presence of the krill *Euphausia superba. The Auk,* 102:540–549.

Offredo, C., V. Ridoux and M. R. Clarke. 1985. Cephalopods in the diets of Emperor and Adélie penguins in Adélie Land, Antarctica. *Marine Biology,* 86:199–202.

Maher, W. J. 1962. Breeding biology of the snow petrel near Cape Hallett, Antarctica. *Condor,* 64:488–499.

Müller-Schwarze, D. 1984. *The Behavior of Penguins Adapted to Ice and Tropics.* Albany: State University of New York Press.

Murphy, R. C. 1936. *Oceanic Birds of South America.* Vols. 1 and 2. New York: American Museum of Natural History.

Parmelee, D. F. 1977. Adaptations of Arctic terns and Antarctic terns within Antarctic ecosystems. In G. A. Llano, ed., *Adaptations Within Antarctic Ecosystems,* pp. 687–702. Washington, D. C.: Smithsonian Institution.

Parmelee, D. F., W. R. Fraser, and D. R. Neilson. 1977. Birds of the Palmer Station area. *Antarctic Journal of the United States,* 12:15–21.

Stonehouse, B. 1970. Adaptation in polar and subpolar penguins (Spheniscidae). In M. W. Holdgate, ed., *Antarctic Ecology,* 1:526–541. New York: Academic Press.

Stonehouse, B., ed. 1975. *The Biology of Penguins.* London: Macmillan.

Volkman, N. J. and W. Trivelpiece. 1981. Nest-site selection among Adélie, chinstrap and gentoo penguins in mixed species rookeries. *Wilson Bulletin,* 93:243–248.

Voous, K. H. 1968. Antarctic birds. In J. van Mieghem and P. van Oye, eds., *Biogeography and Ecology in Antarctica,* pp. 649–690. The Hague: Junk.

Warham, J., R. Watts, and R. J. Dainty. 1976. The composition, energy content and function of the stomach oils of petrels (Order Procellariiformes). *Journal of Experimental Marine Biology and Ecology,* 23:1–13.

Watson, G. E. 1975. *Birds of the Antarctic and Subantarctic.* Washington, D.C.: American Geophysical Union.

Watson, G. E., J. P. Angle, P. C. Harper, M. A. Bridge, R. P. Schlatter, W. L. N. Tickell, J. C. Boyd, and M. M. Boyd. 1971. Birds of the Antarctic and Subantarctic. *Antarctic Map Folio Series,* Folio 14. New York: American Geographical Society.

CHAPTER 7

SECONDARY CONSUMERS— CONSUMERS— MAMMALS

I n addition to birds, fishes, and invertebrates, the Southern Ocean's krill fuel the lives of many marine mammals—whales and seals—that populate this rough, productive, and cold sea. The krill-feeding whales belong to the great suborder Mysticetes. These toothless filterers of sea water strain great quantities of krill through thin plates of horny keratin that hang suspended from the roofs of their enormous mouths. The seals at this trophic level include three species that make a good living in the highest latitudes of the Southern Ocean. This latter group of fascinating mammals will be considered first.

SEALS

The Southern Ocean plays host to six species of seals. Ross seals, southern elephant seals, and Weddell seals live principally on fish and squid and are all secondary and tertiary carnivores. They will be considered with other mammals of their trophic level in chapter 9. Two seal species, the crabeater and Antarctic fur seals, are genuine krill specialists, and are noteworthy in both the biological and economic worlds. Leopard seals split the difference, relying on krill for as much as 50 percent of their diets, and rapaciously attack other seals, penguins, and fish for the rest. This chapter will consider their krill-eating adaptations, and chapter 9 their predatory feats.

Antarctic Fur Seal

Much of the early history of antarctic exploration developed in concert with the quest for the hides of Antarctic fur seals, *Arctocephalus gazella*. This doe-eyed, eared seal was early valued for its luxuriant pelt that became a principal item in the China trade of the nineteenth century. By the end of that era this animal had been ruthlessly exterminated from the subantarctic islands on which it bred in dense colonies. The initial slaughter took place on South Georgia, and, as sealers pushed on in quest of additional breeding colonies, Macquarie Island, the South Shetlands, and other islands were discovered and quickly exploited by them. A Stonington, Connecticut, sealer, Nathanial Palmer, is often credited with being the first human to see the Antarctic Peninsula—which on many maps bears his name.

Although the Antarctic fur seal is not strictly a continental species (it is abundant on islands south of the Antarctic Convergence), it ranges south to the Peninsula, especially now that its numbers are rapidly recovering from the excesses of that nineteenth-century harvest. As a breeding species its colonies are dense and abundant on and around South Georgia, although it also breeds farther south in the South Orkney, South Sandwich, and the South Shetland Islands, not far north of the Peninsula.

Like most land-breeding seals, Antarctic fur seals are highly polygynous; that is, dominant bulls establish territories and defend and breed with several females within them. The competition among males for reproductive rights is probably responsible for the size and sexual dimorphism so often seen in animals with polygynous mating systems. The fur

Antarctic fur seal

Male and female Antarctic fur seal

seals at South Georgia show weights at maturity of 38 kilograms for females (five years old) and 130 kilograms for males (ten years old). Breeding males are 3.5 to 4.5 times as heavy as the females. This inequality expresses itself early in life, for male fur seal pups gain weight faster than females and are significantly heavier by the time of weaning.

The annual breeding cycle of fur seals begins in late October when the mature bulls begin to haul out on the stone and shingle beaches that become their mating territories. These territories initially average about sixty square meters in area. The cows begin to arrive two or three weeks after the first appearance of the bulls and occupy the territories, ten or so cows to a bull. Only two days after hauling out each cow gives birth to a single pup.

Most of the pups are born in any given colony in a short period of time, as many as 80 percent of the young appearing within a 17-day period. Seven or eight days after parturition the cow comes into estrous and mates with the bull of her harem.

During this period the bulls continue to jockey for territory and the cows that occupy it. The early arrivals stake out claims nearest the water's edge (but safely above the storm tide mark). Tardy bulls must be content with more inland sites, and must run a gauntlet through established territories to get to these. Initially the amount of fighting among bulls is small, but as the density of the population increases with the later arrivals, more and more aggression occurs. The bulls of adjacent territories defend their space and females with displays that involve oblique stares with the head held high. Little overt fighting occurs unless a strange bull blunders into an established territory. Even so the territories diminish in extent, averaging only little more than 20 square meters at the height of the breeding season.

After about five weeks the bulls have finished mating and are ready to return to the sea for a feeding binge. Meanwhile, the cows repeatedly return to the water for three- to six-day forays, nursing their young for two- to five-day periods between the feeding trips. These feeding excursions begin right after mating, when the pups are but eight days old. In all, about 17 feeding forays are made by the mother during her lactation, each of an average duration of four days.

The intermittent feeding of fur seal pups has consequences that mark the breeding biology of these seals as special. First of all, the number of pups on the beach at any given time greatly outnumbers the number of adults. Because these pups are not fed regularly their growth rates are slower than those of other species of seals, and the time to weaning is correspondingly longer. At Bird Island, near South Georgia, weaning does not occur until early March for pups born at the end of November.

The recovery of Antarctic fur seal populations from the near extinction that occurred in the nineteenth century has been remarkable. As recently as the 1930s only a relict population of a hundred or so animals was known at Bird Island. By 1982 estimates of the total population were as high as 900,000 animals, with an annual pup production of perhaps 250,000. Based on this current rate of population increase, as many as four million Antarctic fur seals could exist by the year 2000. New breeding colonies of fur seals are established every year as the exploding population spreads away from its South Georgia focus. As breeding beaches become crowded, the behavioral interactions between fur seals have more dire consequences. Not only are conflicts between bulls more frequent and severe, but the mortality of pups rises as a direct consequence of population density.

At low-density, newly colonized beaches, only 3 to 6 percent of the pups die. At sites of high-seal-density pup mortality involves as many as 31 percent of the pre-weaning animals. The causes of this mortality are mainly starvation and severe skull damage. Starvation seems to occur because the large number of pups that appear in a short period of time apparently confuses the new mothers to the point that they do not learn (imprint) their newborn effectively. A young pup that can't be recognized by its mother is doomed—either to starve, or to be attacked when it attempts to suckle from an unrelated cow. The high incidence of skull injuries (28 percent of the total mortality in the dense population) is due mainly to females biting strange pups that attempt to suckle. A further cause of mortality (11 percent of the total) is due to ruptured pup livers—caused when they are inadvertently trampled or rolled on by large bulls.

Antarctic fur seal pups weigh a little over six kilograms at birth. Despite their irregular feeding schedules they gain about 80 grams a day to reach an average weaning weight of almost 16 kilograms after 110 days of nursing.

Of the food provided to the pup by its mother only about 15 percent is channeled to new growth, the remainder being spent for the metabolic requirements of the pup. A ten kilogram weight gain before weaning thus represents something on the order of 125 kilograms of food eaten by the mother—just to nourish her young.

If a female fur seal completes 17 feeding trips during her period of lactation, each one must result in the capture of more than 7 kilograms of krill that will result in milk production. The krill that are captured by these seals near South Georgia are predominantly mature females with an average individual weight of about 1.1 grams. The arithmetic suggests that each cow must capture more than 6,000 krill per trip in order to feed her young. The cow's own metabolic requirements demand an even larger krill consumption.

Although krill make up the bulk of the diet of nursing fur seal mothers, they also eat other sorts of food, particularly outside of the breeding season. Some studies of stomach contents show a diet that consists of one-third krill, one-third squid, and one-third fish. Indeed, it is characteristic of these seals to have teeth stained and blackened from squid ink.

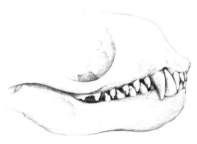

Jaws and teeth of Antarctic fur seal

Following breeding the bulls return to the ocean and go through a molt that is not nearly as catastrophic as the molt suffered, for instance, by the southern elephant seals, which shed both hair and epidermis and cannot enter the water during this period. The cow molts piecemeal during her lactation. The reproductive season concludes by early May when the last cows and pups vacate the breeding beaches.

Age determination through the examination of growth rings in the teeth suggest that this seal lives to a maximum age of about 30 years for the female. Cows first bear pups at age three, and about 50 percent of the females born survive to this age.

The biology of non-breeding Antarctic fur seals is poorly known. Some have been sighted off the coast of Patagonia, and non-breeding animals are common near the Antarctic Peninsula. A few tag recoveries suggest that they range up to 2,000 kilometers from the breeding islands. Despite the past economic importance of this animal, and our considerable knowledge of its reproductive biology, much remains to be learned of its total life history.

Growth rings in a sectioned seal tooth

Crabeater Seal

If the Antarctic fur seal represents a successful story of recovery from overzealous exploitation at the hand of man, the crabeater seal, *Lobodon*

Crabeater seal

carcinophagus, presents a different example of how our species' insatiable search for economic gain can affect the natural world. At the present time, and for the forseeable future, this species of phocid seal is the most abundant seal on earth. Of the thirty-six species of living pinnipeds, crabeaters are estimated to account for more than 56 percent of the worldwide total number of seals. And, because they are one of the larger species (adults average about 193 kilograms body weight), they account for about 79 percent of the total seal biomass. Moreover, the population of this seal seems to be increasing at a steady rate. These eye-opening statistics are especially impressive considering that this animal lives in an environment, the antarctic pack ice, that seems to us to be particularly inhospitable.

In size, shape, and behavior, crabeater seals are among the most engaging and pleasing of Antarctic seals. They are streamlined, have pointed snouts, and are variously colored from cream to silver, with the newly molted animals (the molt occurs in January), being the lightest in coloration. A moderate amount of flecking is seen on the shoulders, sides, and bellies of some individuals. Leopard seal scars are often evident.

Unlike fur seals, crabeaters are not gregarious. Except for aggregations of juveniles that involve a few hundred or so individuals, most crabeaters are encountered as solitary adults, breeding pairs, or breeding pairs with a pup. They practice a different mating system than do fur seals, with males defending but a single female. A consequence of this behavior is that the pronounced sexual dimorphism of fur seals is absent and, to some extent, is reversed in crabeaters. Any difference in size between the sexes results from the fact that female crabeaters continue to grow be-

Scars on a crabeater seal

yond the point (at about age seven) when males stop. At seven both sexes weigh about 200 kilograms. Males maintain this weight, but older females continue to gain slowly, so that the average 15-year-old female crabeater weighs about 215 kilograms.

Like fur seals, crabeaters are most conspicuous during their reproductive season. Unlike fur seals, however, their entire "terrestrial" existence is spent on ice floes, not on land. Their biology, therefore, is less well-known.

During the breeding season, which occurs in October at the pack ice fringe where these seals are most abundant, the usual sighting of crabeaters encompasses "family" groups that consist of a mature male, a female that usually gives birth about midway through the month, and (after parturition) her newborn pup. The ice floe chosen for this event is of modest size and is hummocked, providing both protection from wind and a vantage point from which the male can survey the seascape. The male is vigilant and defends a territory of about 50 meters radius against other males, should they try to enter. Males do displace one another from thee territories, but the frequency with which this happens is not known.

The female nurses her pup for a period of about four weeks, while it

"Family" group of crabeaters on ice floe

grows from a birth weight of about 26 kilograms to a weaning weight of 110 kilograms. She does not eat during this period, and loses about half of her body weight.

In four or five weeks, when the pup has molted its lanugo fur and is ready for weaning, its mother comes into estrous. Prior to this time the attending male made sexual advances that were rebuffed by the female biting his head and neck (older male crabeaters are much scarred about the head). As ovulation approaches the male's advances become more determined and the female suffers courtship bites on her neck and back. This behavior is so intense that the estral female is literally covered with her own blood.

It is not certain if mating occurs on ice or in water. Because crabeaters are found in areas frequented by enemies such as killer whales and leopard seals, mating on the ice could take place. Cryophilic seals like Weddell seals that breed in fast ice regions where predators usually do not occur, normally mate under water.

After mating, the male forsakes the female and searches (probably by smell) for another pre-estral cow. Although the groupings of these seals are termed "family" there is no evidence that the male in attendance is, in fact, the father of the pup of the year. The failure of crabeater seals to show social organization during the remainder of the year suggests that permanent pair bonds are probably not established.

Although fertilization of the egg that will produce next year's pup occurs soon after mating, the development of that embryo is arrested in the blastocyst stage, before implantation. This sort of suspended animation continues for a period of two or three months until well after the present year's pup is weaned. The effect is to produce a gestation period of almost one year, although the embryo is actually developing for a shorter period of time than this.

Lactation in crabeaters is shorter and more intense than that of fur seals, and the growth of the pups continues at a rapid rate after weaning. Sexual maturity is reached in as long as six years, or in as little as two years, depending on the status of the food supply and other biological factors. During their first year of life young crabeaters are particularly susceptible to predation by leopard seals. By the end of their first year, however, they reach a body size that makes them less vulnerable, and escapes from leopard seal attacks become so common that most crabeaters carry some scars inflicted by these predators.

Crabeaters typically haul out at midday when the sun is near its zenith, and on the ice they are the most alert and wary seals found in the Antarctic. Humans approach them closely with greater difficulty than the more sedentary Weddell seal—partly because of their more nervous behavior and partly because of the inaccessibility of their habitat. Consequently we know less of their physiology and general biology.

Crabeaters feed most intensively in the dusk, dawn, and nocturnal hours. It is at this time of day that the krill which they eat rise closest to the surface. The duration of feeding dives, which average as much as ten minutes, suggest that they may possibly dive as deeply as Weddell seals or southern elephant seals. Their usual dives, however, are probably only to depths of 50 to 80 meters, although recent measurements show they can descend to as much as 400 meters.

Stomach contents of crabeaters prove that they specialize on krill. Ninety-four percent of the food items in their stomachs are euphausiids, with fish (3 percent) and squid (2 percent) taken in small quantities. Rocks and sand are frequently found in the stomachs too. For the purpose of krill consumption crabeaters are equipped with one of the most unusual dentitions among mammals. The molar and premolar teeth of these animals have several elongated cusps, and the teeth of the upper and lower jaws occlude so a sieve is formed that efficiently strains krill from water brought into the mouth. There is even a bony knob behind the last molar that effectively closes a potential "escape" route for krill.

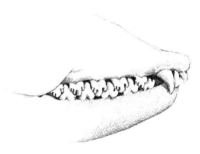

Jaws and teeth of crabeater seal

Two methods of feeding have been described for crabeaters. In the first, the seals swim through concentrations of krill with the lower jaw held widely open, filtering water in a method analogous to that used by filter-feeding sharks and Right whales. But crabeaters have also been observed to suck bottom invertebrates into the mouth through pursed and extended lips.

Whatever the feeding mechanism, these seals eat prodigious amounts of krill. Based on an analysis of their energetic requirements, an average-size adult crabeater of 193 kilograms requires a little less than 15 kilograms of krill per day. Because there are periods (breeding and molting) when either feeding does not occur or is reduced, the seal's compensatory feeding has it eating closer to 20 kilograms per day. Stomach contents of about 8 kilograms have been measured, so it seems reasonable to think that they may feed to satiation at least twice a day when food is available.

Some antarctic biologists argue that food for the crabeaters has become more available in recent years with the demise of the former vast populations of baleen whales. With fewer whales to crop krill production, a greater food supply has been channeled to other krill-eaters like seals, sea birds, and fish. A consequence for the crabeater has been an increase in their abundance. Already a major presence in the pack ice fringe around Antarctica, the numbers of crabeaters are increasing rapidly. The present population is conservatively estimated to be not much less than 20 million animals, and is projected by some to reach 50 million by the year 2000.

Not only has the population of these animals grown, but they are responding to the glut of food by reaching sexual maturity faster. Back calculations of age and growth from tooth samples indicate that the average age of maturity in females has dropped from nearly 4 years to 2.5 years; in the period of time that began with the cessation of large-scale whaling operations.

If crabeater seals present a life of specialization for feeding and breeding in a food-rich but physically hostile situation, their combined competitor and predator, the leopard seal, shows a more diverse set of adaptations to the same environment.

Leopard Seal

By nearly all criteria the leopard seal, *Hydrurga leptonyx*, is one of the most interesting animals found in the Peninsula region. This elongate seal (large females approach 4 meters in length and 450 kilograms in weight—males are somewhat smaller) ranges further, feeds more diversely, and exists in smaller numbers than most other seals of the Southern Ocean. They are usually encountered as single individuals, except when a female is nursing her pup. Added to the problems of low population density and its lack of gregariousness are the vigilance of this seal and its ferocious underwater predatory behavior. All of these factors conspire to prevent much detail in our understanding of most aspects of its biology.

The leopard seal is a sleek, trim animal, with an elongate neck and a characteristically large head that ends in a pointed snout. The exceptionally large mouth is tipped with strong canine teeth in front of complex, tri-lobed molars that resemble those of the crabeater. The combination of a long, snake-like neck, triangular head, and prominent teeth results in the frequent description of the leopard seal head as "reptilian." Their common name, however, derives both from the spotted character of their

Leopard seal

coats and from their solitary hunting behavior that sometimes features the ambush of favored prey such as penguins.

When hauled out on pack ice this seal displays a coat that is dark above and silvery below. The black spots on the throat are particularly prominent when the seal characteristically raises its head to stare directly at an approaching intruder. They impress scientists studying them with their "curiosity" of human intruders. Unlike other seals of the pack ice leopards hold their foreflippers close to the body when on ice and can move quickly to the water with "porpoising" motions of the body—which they do promptly upon even slight provocation.

Although leopard seals are found in greatest abundance in the fringe of the pack ice area, they also range well to the north (especially in winter), and are frequent visitors to subantarctic islands and even the South American mainland. Their abundance at places such as Macquarie Island seem cyclical and probably relate to factors that influence their food supply.

Leopard seals have the most catholic diet of antarctic pinnepeds. Stomach analyses put krill as their number one prey (especially in winter), with these crustaceans making up about 50 percent of the diet. But leopard seals also eat penguins (20 percent), other seals (14 percent), fish (9 percent), and cephalopods (6 percent). Penguins and seals become most important during the spring and summer breeding season when chicks and pups fall prey to the leopards. The range and variety of the diet suggests a suite of adaptations to deal with these different food types.

The most important element of their diet, krill, is clearly the object of one major feeding adaptation, the molar teeth. With the jaws closed, the occluded molars interdigitate in much the same manner as in the crabeater seal and form an effective sieve to retain krill in the mouth. When feeding on krill, leopard seals dive to comparatively shallow depths, probably not much deeper than 50 to 60 meters, and average a little more than 2.6 minutes under water. Like crabeaters their feeding activity is concentrated during the crepuscular and nocturnal hours when the krill rise toward the surface.

Jaws and teeth of leopard seal

But leopard seals are also capable of predatory feats of considerable magnitude, particularly when they choose to feed on penguins and other seals—especially crabeater seals. There is evidence that younger leopard seals confine their feeding activities to krill, and later, as they grow older, they switch more and more to larger prey. This may be particularly true of males, who are active predators of young seals and penguins during the period when the females are pupping. The activities of these hunting leopard seals will be covered in chapter 9.

Not much solid information exists concerning the reproductive biology of leopard seals. Courtship and mating probably occur under water, most likely following the weaning of the pup of the previous reproductive season. Near the Peninsula pupping occurs during November and December. The young are born on moderate-sized ice floes with only the mother in attendance. Nursing is intense and probably does not last for more than about one month. The females, which grow to a mature body size almost 10 percent larger than the males (average adult body weight for the species is about 315 kilograms), present one of the seal world's most extreme cases of female-larger sex dimorphism. The evolutionary causes and meaning of this are not agreed upon, although smaller and more agile males could have some advantage in an underwater mating system.

As in other antarctic seals, sexual maturity is probably reached in five years or less, and the longevity of individual seals stretches for fifteen years or longer.

Population estimates, based on counts of leopard seals hauled out during their favorite basking hours hear midday suggest that in 1982 approximately 440,000 individuals existed in the Southern Ocean. That number seems to be increasing at the present time, and by the year 2000 as many as 600,000 leopard seals may be with us.

Although leopard seals appear asocial, they make sounds of a sort not heard from the silent crabeater seal. These vocalizations are made both above and in the water and have a pulsed character (300 per second). People who have heard them remark on their "haunting" quality. Their meaning, whether in communication, echo-location, or hunting, is not known.

WHALES

Although the seals that feast on the abundant krill of the Southern Ocean are exceedingly important elements of this ecosystem, most people are more impressed with the history and biology of the other great mammalian krill-eaters of this region—the whales.

The Southern Ocean sets its table to a host of cetaceans. In all, about 15 species have been recorded from the waters below 60° South. But of these, only half are baleen whales (Mysticetes) that derive their nourishment from herbivores such as krill.

The Mysticete whales of Antarctica are fabulous in their biology and in the history of fisheries that developed for them. The success of those fisheries in this century resulted in the decimation of the baleen whale stocks in the Southern Ocean. That decline created ecological waves that continue to reverberate throughout the Southern Hemisphere and beyond and will be considered in chapter 10.

There exist in the Southern Ocean seven species or subspecies of baleen whales. The overriding characteristic of these animals is the presence of

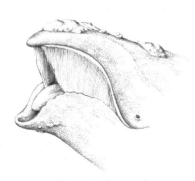

Mouth of rorqual

slender plates of horny material (keratin) that hang in two festooned rows from the sides of the roofs of their mouths. These mats of baleen form a filter through which sea water with its included plankton can be strained to separate the plankton and prepare it for being swallowed.

The antarctic whales that make their living this way include the Minke whale *(Balaenoptera acutorostrata),* Blue whale *(B. musculus),* Pygmy Blue whale *(B. musculus brevicauda),* Fin whale *(B. physalus),* Sei whale *(B. borealis),* Humpback whale *(Megaptera novaeangliae),* and Southern Right whale *(Eubalaena australis).* The last named species has the finest and most elaborate baleen, and therefore is capable of filtering the smallest plankton for its food. Fished to near extinction, the Southern right whale is now recovering, but because its normal range is well to the north of the Antarctic Peninsula, it is not a species well within the scope of this book.

In a similar manner, Sei whales, which also are able (especially as juveniles and subadults) to eat the smaller sizes of plankton, visit the northern reaches of the Southern Ocean near the Antarctic Convergence, and seldom venture near the Peninsula. Pygmy Blue whales, a recently described subspecies, are known only from the eastern sectors of the Southern Ocean. They will not be considered here.

The remaining baleen whales, Blue, Fin, Minke, and Humpback, are all to be found in the waters adjacent to the Antarctic Peninsula and are worth our attention. With the exception of the Minke whale, there is a consistent life history strategy in these great animals. All share the pattern of leaving the Southern Ocean in the winter and migrating to lower latitudes where warmer waters favor calving, mating, and reduced metabolism during the time of the year when krill are too deep in the Southern Ocean to be accessible.

During their winter sojourn to the north, the antarctic whales feed very little for the types and densities of plankton available to them above the Antarctic Convergence are not favorable. Consequently the whales go through a period of semi-starvation, living off fat stored from their antarctic feeding binge. When the whales return to the Southern Ocean for the austral summer, their blubber is reduced in thickness, and their condition is poor. Even though their food intake is minimal on the wintering grounds (which possibly extend to the equator or beyond in some species), estimates of their metabolic expenditures suggest that without the small amount of food they do manage to find, the fat that they store during the summer might not see them through a six-month period of starvation. Reduced as it is, the winter diet must be critical for the survival of these animals.

All of these rorquals (the common name given to balaenopterid whales) have populations in the Northern Hemisphere. Even though the northern populations move south during the boreal winter, there is probably no exchange, and the northern and southern stocks are genetically isolated from one another.

In their migrations to and from the Southern Ocean, species such as the Blue whale show a regular separation of sex and age classes. The first whales to move south in the spring (and the last to return in the fall) are the pregnant females. Spending more than six months in the krill-rich waters off the Peninsula, these females build up maximum blubber and fat stores to prepare for the metabolic rigors of birth and lactation. Because female rorquals are larger than males, these pregnant Blue whales were the heaviest and richest catches in the bygone whaling industry.

The next class of migrants includes the mature males that are followed, in turn, by immature males and females, and finally the lactating females with their newly delivered calves. The order of this migratory sequence is reversed during the northward return in the fall. Then the lactating females leave first, followed progressively by immatures, mature males, and finally, the pregnant females.

Although rorquals migrate in loose aggregations of a few individuals, and show "patchiness" in their distribution during the summer feeding season, the only apparently intense social bond is that between mother and nursing calf. These filter-feeding Mysticetes have little in the way of complex social organization.

The several species of baleen whales penetrate the Southern Ocean to different degrees of latitude. This, plus the staging of their migrations and their failure to form strong social units, reduces direct competition for the krill that all of them feast on. The Minke whale moves right to the pack ice edge and beyond. It, uniquely among the rorquals, occasionally overwinters in the Antarctic. The Blue whale journeys almost as far south as the Minke. Fin whales and Humpbacks summer in latitudes of the Peninsula, while the Sei and Southern Right whales stay pretty much in the northern part of the Southern Ocean.

Minke Whale

Minke whales *(Balaenoptera acutorostrata)* are distinctive both for their size and their relative abundance in the Antarctic. The smallest of the Mysticetes, Minkes mature at about 7.5 meters and grow to a maximum length of only 9 meters. They are the most numerous baleen whale in the Southern Ocean, with present-day population estimates ranging from 173,600 to 323,500 individuals. Because of their small size and the small amount of blubber that they contain (averaging only 15 percent of their body weight) they were not sought out by whalers so long as the populations of larger rorquals held out. That has changed today.

Minke whale near ice

Minkes are distinctive for more than their abundance and small size. More than any other baleen whale, they penetrate right to the ice edge where they feed among the rich surface waters for the krill (both *Euphausia superba* and *E. crystallorophias)* that almost exclusively make up their diets in the Southern Hemisphere. It is worth noting that Minke whales are also abundant in the northern hemisphere where they often feed on small fishes such as capelin and herring.

Another distinction of the Minke whale is its reproductive cycle. Females bear young on an annual cycle, with mating following parturition. Compared to other rorquals the percentage of females that are simultaneously lactating and pregnant is high. The reproductive potential of this species is thus much higher than the theoretical limits of larger whales. Although this greater reproductive competency could result in more rapidly recovering populations, the smaller size of the Minke makes it more vulnerable to the predatory advances of Killer whales (chapter 9).

Blue Whale

Surely one of the most fabulous creatures the world has ever seen, the Blue whale *(Balaenopterus musculus)* is the all-time winner for size. These immense animals mature at 23 meters and reach a maximum size of at least 30.5 meters. At the final length, a Blue whale in prime condition at the end of the summer feeding season could weigh between 150 and 200 tons, although the average is closer to 100 tons. At these lengths and weights they are the largest species of animal that ever breathed on earth, dwarfing even the largest dinosaur.

The secrets of the Blue whale's success are krill, and, in concert with

Blue whale and Minke whale

Pleated throat of a rorqual

the other rorquals, an efficient means of cropping them. As the swarms of *E. superba* rise in the summer to the surface of the Southern Ocean they come within reach of the unique feeding mechanism of these whales. Unlike the filter feeding sharks (whale, basking, and manta rays) and Right whales that open their mouths and ram filter plankton as they swim through the water, the rorquals take mouthfuls of krill-rich water. The enormous mouth and pharynx of the whale is remarkably distensible due, in part, to longitudinal folds or grooves that run from under the chin, well back into the abdominal region.

Having engulfed an enormous mouthful of krill-laden water, the Blue whale then uses its huge, piston-like tongue to drive the water through the two dense rows of baleen plates that hang inside of each upper jaw. Krill and other small organisms are trapped in the meshes of the baleen and are then worked to the back of the throat for swallowing before the whale bites the ocean again.

Each row of baleen is composed of a series of triangular plates that can be as long as 2 meters in a Right whale and almost a meter wide across

Baleen plate

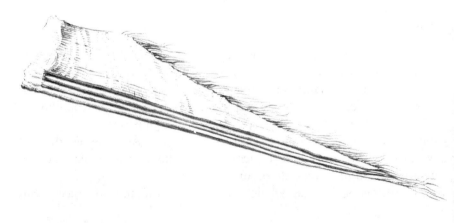

the base. These plates (each about 5 millimeters thick) are placed transversely with about 10 millimeters separating each plate. The inner edges of the plates become worn with use and their central tubules or filaments are exposed like the frayed ends of ropes. These fine strands form the mesh of the filter. When the whale's enormous tongue forces water through the mat of fine strands of frayed baleen, the plankton are strained out and then swallowed.

To some degree the size of the baleen strands determines the size of the prey which can become enmeshed in them. Right whales, having the finest baleen can (and do) subsist on a diet of relatively small copepods. Sei whales also are known for their fine baleen and consequently the range of food items that appear in their diets. The coarser baleen of the high Antarctic whales (Minke, Blue, and Fin) restrict them to the larger adult euphausiids from 27 to 60 millimeters long.

These whales eat prodigious amounts of krill. Fully 80 to 100 percent of the whales captured in the old antarctic whaling industry were found to have food in their stomachs, suggesting that they eat two to five meals a day. At their maximum feeding intensity, the great rorquals ingest 3 or 4 percent of their body weights per day. For a 100-ton Blue whale that means three or four tons of krill per day (at one gram per krill, this works out to about three or four million krill per day). Clearly, it may be advantageous for these whales to keep their distance from one another and not gather in dense social groups while feeding.

Because of the old "fishery" for Blue whales, good information is available for the migratory patterns of these animals as they came to and departed from the whaling grounds. They conform completely to the general pattern given above. That is, pregnant females arrive first, followed by males, immatures, and lactating females.

The early antarctic whalers were intrigued by the newly arrived whales, which could be identified both by their relative thinness and the presence of numerous small circular gouges carved out of their skins. Now we know that these gouges were the work of a curious deep water pelagic elasmobranch, the cookie-cutter shark *(Isistius brasiliensis)*. This little fish (big ones are only 30 centimeters long) reacts to a large fish or a whale passing close by with an attack and a spinning wrench that enables it to carve out a plug of skin and blubber or meat from the prey. There is irony in the fact that the largest creature ever to swim is not completely immune to predation by one of the smaller sharks.

Fin Whale

The second largest of the rorquals (and thus of all living whales) is the Fin whale *(Balaenoptera physalus)*, which matures at 19.5 meters and grows

A Fin whale breaching

to at least 26 meters. Slimmer than the Blue Whale, Fin whales reach a maximum weight of between 80 and 90 tons, which is close to the heaviest Right whales—a shorter, more robust species. In the Northern hemisphere, Fin whales are known for their agility, rolling over on their sides and even snapping sideways at significant concentrations of prey. Baleen whales are responsive to tactile contact with their planktonic prey, and have whiskers or vibrissae that are probably important in their feeding behavior.

The reproductive biology of Fin whales is representative of the other rorquals, and presents a number of mysteries to whale biologists. Mating takes place on the wintering grounds at low latitudes. Actually, the breeding grounds for most of the Southern Hemisphere whales are not well-known as few records of catches in the northern end of the range exist.

Male rorquals show periods of spermatogenesis that imply a seasonal nature to their mating activities. Were a female to ovulate outside of the proper season, she probably could not find a mate. The details of mate selection and courtship are sketchy at best. Monogamy is thought to exist, although it is not known if mated pairs are faithful from one reproductive season to the next. Some baleen whales (Humpback, for example) are known for vocalizations ("songs") that can be transmitted for hundreds of kilometers underwater. The meaning of these sounds is debated, but a likely function can be the advertisement of mating intentions. Vocalizations in the other baleen whales are not so well studied.

Activities associated with mating are reliably reported only for Humpback whales, and accounts vary between vertical and horizontal copulatory positions. The detailed structure of the penis of male rorquals and the few reported observations argue for a brief mating act, lasting but a few seconds. Courtship, it might be expected, lasts considerably longer.

Having conceived, the female Fin whale begins an 11-month period of pregnancy that results in the birthing of her calf on the sparse winter grounds. At this time she is in the peak of condition, having spent the preceding six months in the rich Southern Ocean. The newborn calf of the Fin whale probably is about 6.4 meters long at birth and weighs about one ton. Blue whale calves are between 7 and 8.2 meters and could weigh 3 tons at birth.

An enormous amount of energy goes into nursing the newborn calf. Blue whale calves grow at a rate of nearly four centimeters per day until weaning at seven months of age. By this time they are more than 16 meters long and weigh 23 tons. That amounts to average weight gains of more than 80 kilograms per day throughout lactation. The amount of milk (which is extraordinarily rich in fats) produced by the mother must exceed 130 kilograms per day. It is no wonder that a lactating cow loses nearly half of her body weight before reaching the feeding areas of the Southern Ocean in the early summer.

While the growth of Fin whale calves is not quite as impressive as that of the Blue whale, it is proportional. Where the Blue whales are the largest animals ever to have lived, Fin whales rank second. Fin whales grow at more than 52 kilograms per day as nursing calves—no weight gain to sneeze at! They wean at a length of about 11.5 meters.

Determining the age of rorquals is an uncertain task but one which is

critical to the proper management of their stocks. Historically these animals were judged to have maintained their preweaning growth rates into their subadult lives. Such projections estimated sexual maturity in only two years. It turns out, however, that growth slows with weaning, and maturity (based on annuli laid down in the ear plugs that develop in each external auditory meatus) is not reached until the whales are six to ten years old. They might have lived to be 20 or 40 years old in a pristine world that did not have whalers.

One consequence of the early overestimation of growth rates was a failure to establish controls on the slaughter of these great animals until too late. Stocks of the Fin and Blue whales are now mere pittances of their former numbers. The decline in populations from the early years of this century until now is estimated to be from 400,000 to 85,200 Fin whales; and from 180,000 to 8,000 Blue whales. Some consequences of this decline will be considered in chapter 10.

Humpback Whale

The Humpback whale *(Megaptera novaeangliae)* differs from other rorquals in several critical features. It is not as long as most (15.2 meters maximum), but it is heavy for its length (60 to 70 tons maximum). Easily distinguished in the field by its huge fore flippers (from which its generic name is derived), it is more likely to be infested with barnacles than are the other rorquals. Most important, however, is the fact that this whale prefers inshore waters, over continental shelves, for its feeding areas and migration routes. The other rorquals stay farther out to sea.

This last feature of its biology made this whale, alone of the giant baleen whales, susceptible to shore-based whale fisheries in both the Northern and Southern Hemispheres. Indeed, Humpback populations in the early years of this century had been decimated before the pelagic fisheries for Blue and Fin whales even began.

The ancestral migration routes of the southern population of Humpback whales take them along both the east and west coasts of South America between the Southern Ocean and their subtropical wintering and breeding areas. Both migration routes funnel them south to the Antarctic Peninsula where these whales spend the summer feeding in the Bellingshausen Sea just to the west of the Peninsula.

In the Northern Hemisphere Humpbacks feed heavily on fishes such as capelin, herring, and sand lance. They use several feeding behaviors to concentrate schools of these fishes, which they then engulf. The antarctic Humpbacks, on the other hand, feed exclusively on krill, taking as much as 500 kilograms of euphausiids at each of four meals a day. A single Humpback thus can eat two tons of krill a day, and 240 tons over a full season.

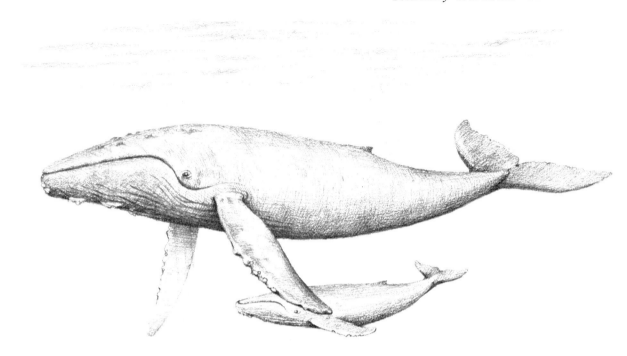

Humpback whale and calf

Estimates of the total initial population of Humpbacks suggest that around 100,000 of them may have visited the Southern Ocean each year before whaling began. Today's population is not much more than 3,000 whales. The difference, 97,000 whales, represents 23,280,000 tons of krill that were once harvested by these whales, but are not today.

Having discussed the immediate fate of krill and other primary consumers in the Antarctic marine ecosystem, it is now possible to turn to the next tropic level—that of the tertiary consumers (secondary predators). With each trophic level the amount of energy available to fuel life decreases by about an order of magnitude. That fact is expressed in both the smaller populations of higher predators, and in the extreme adaptations that limited energy requires of them. Here, life gets more interesting. These points of view will be illustrated with a look at predatory birds of the Antarctic (chapter 8), and then the mammals that share this trophic level (chapter 9).

ADDITIONAL READING

Bertrand, K. J. 1971. *Americans in Antarctica.* New York: American Geophysical Society.

Brown, S. G. and C. H. Lockyer. 1984. Whales. In R. M. Laws, ed., *Antarctic Ecology,* 2:717–781. London and Orlando, Fl.: Academic Press.

Croxall, J. P. and M. N. Pilcher. 1984. Characteristics of krill *Euphausia superba* eaten by Antarctic fur seals *Arctocephalus gazella* at South Georgia. *British Antarctic Survey Bulletin,* 63:117–125.

Doidge, D. W., J. P. Croxall and J. R. Baker. 1984. Density-dependent pup mortality in the Antarctic fur seal *Arctocephalus gazella* at South Georgia. *Journal of Zoology* (London), 202:449–460.

Doidge, D. W., J. P. Croxall, and C. Ricketts. 1984. Growth rates of Antarctic fur seal *Arctocephalus gazella* pups at South Georgia. *Journal of Zoology* (London), 203:87–93.

Gaskin, D. E. 1982. *The Ecology of Whales and Dolphins.* London and Exeter, N.H.: Heinemann.

Goodall, R. N. P. and A. R. Galeazzi. 1985. A review of the food habits of the small cetaceans of the Antarctic and Subantarctic. In W. R. Siegfried, P. R. Condy, and R. M. Laws, eds., *Antarctic Nutrient Cycles and Food Webs,* pp. 566–572. Berlin: Springer Verlag.

Hain, J. H. W., M. A. M. Hyman, R. D. Kenny, and H. E. Winn. 1985. The role of cetaceans in the shelfedge region of the northeastern United States. *Marine Fisheries Review,* 47(1):13–17.

King, J. E. 1983. *Seals of the World.* 2d ed. Ithaca, N.Y.: Cornell University Press.

Laws, R. M. 1977. The significance of vertebrates in the Antarctic marine ecosystem. In G. A. Llano, ed., *Adaptations Within Antarctic Ecosystems,* pp. 411–438. Washington, D.C.: Smithsonian Institution.

Laws, R. M. 1984. Seals. In R. M. Laws, ed., *Antarctic Ecology,* 2:621–785. London and Orlando, Fl.: Academic Press.

Laws, R. M. 1985. The ecology of the Southern Ocean. *American Scientist,* 73:26–40.

Mackintosh, N. A. 1970. Whales and krill in the twentieth century. In M. W. Holdgate, ed., *Antarctic Ecology,* 1:195–212. London and New York: Academic Press.

Matthews, L. H. 1978. *The Natural History of the Whale.* New York: Columbia University Press.

Oritsland, T. 1970. Sealing and seal research in the south-west Atlantic pack ice, Sept.–Oct. 1964. In M. W. Holdgate, ed., *Antarctic Ecology,* 1:367–376. London and New York: Academic Press.

Ray, C. 1970. Population ecology of Antarctic seals. In M. W. Holdgate, ed., *Antarctic Ecology,* 1:398–414. London and New York: Academic Press.

Ridgway, S. H. and R. J. Harrison, eds. 1981. *Handbook of Marine Mammals.* Vol. 1: *The Walrus, Sea Lions, Fur Seal and Sea Otter;* Vol. 2: *Seals.* New York: Academic Press.

Siniff, D. B., I. Stirling, J. L. Bengston, and R. A. Reichle. 1979. Social and reproductive behavior of crabeater seals *(Lobodon carcinophagus)* during the austral spring. *Canadian Journal of Zoology,* 57:2,243–2,255.

Siniff, D. B. and S. Stone. 1985. The role of the leopard seal in the tropho-dynamics of the Antarctic marine system. In W. R. Siegfried, P. R. Condy, and R. M. Laws, eds., *Antarctic Nutrient Cycles and Food Webs,* pp. 555–560. Berlin: Springer Verlag.

Slijper, E. J. 1979. *Whales.* 2d ed. Ithaca, N.Y.: Cornell University Press.

Small, G. L. 1971. *The Blue Whale.* New York: Columbia University Press.

Zenkovich, B. A. 1970. Whales and plankton in Antarctic waters. In M. W. Holdgate, ed., *Antarctic Ecology,* 1:183–185. London and New York: Academic Press.

CHAPTER 8

SECONDARY PREDATORS— BIRDS

When krill and other zooplankton are eaten by primary carnivores, energy is diminished in the ecological sense. The fishes, squids, whales, penguins, and other sea birds that feast on the Southern Ocean's abundant herbivores are converted, in part, into new growth. But this part is small indeed. A vast percentage of that energy is lost to the living antarctic community both as waste heat generated by metabolism, and the entropy dictated by the grim thermodynamic reaper. Although the efficiencies of energy conversion in the higher trophic levels of the antarctic ecosystem are difficult to measure or estimate, the expectation for better studied systems is that something like 10 percent of the food energy available to secondary carnivores will be converted into new biomass.

The constant cold environment of high latitudes extracts a metabolic price from the endotherms (constant temperature, warm-blooded animals) that do business there. All of these animals have batteries of anatomical, behavioral, and physiological adaptations that allow them to cope with low temperatures. These include: thick insulating layers in the form

of fat (blubber) or dense plumage; adjustments in blood flow patterns that concentrate body heat in the core of the animal; behavioral patterns that involve shelter-seeking in extreme weather and sun-basking in more equable conditions; and even cellular modifications for survival at low body temperatures when the other strategies fail. These elegant adaptations are needed to maintain the thermal gradients necessary for life in Antarctica where the temperature differential from the body core to the outside is nearly always at least 40° C, and may often reach 80° C or more.

Many sea birds live to fairly old ages. Skuas, for example, are known to reach ages exceeding twenty years. Southern Giant Petrels may live to be older than fifty. The point here is that, having reached mature body weight (usually within a year or less), further net production (growth or the addition of new tissue) is confined to the production of eggs and young nourished to the point of weaning or fledging. Consider yourself as an example. Having reached your normal body weight (generally well before your twentieth birthday) you continue to eat a daily average of, let's say, 2,500 kilocalories (a kilocalorie is one dietary calory) for the rest of your life—without, hopefully, gaining further weight. Nearly all of that food energy is used in heat production, normal cell replacement, and molecular repair. The longer you live, your net production (the energy represented by your body weight, plus the birth weight and weaning weight of any children you have borne) becomes a smaller percentage of your total or gross production (all of the food you have eaten since weaning). Thus the older the animals are in a particular trophic level, the more inefficient will be the energy transfer through that trophic level.

There are good reasons why predators live long lives, and we will explore some of them in this and the next chapters. But basically the main problem is that for this and succeeding trophic levels a steady, rich food supply is seldom available. The fact is, the thermodynamic inefficiencies and metabolic heat loss of the foregoing trophic levels have reduced the rich production of the Southern Ocean to a few separate feasts for the secondary predators. Their existence, then, can often be a lean and precarious one. To support this idea, we'll now look at the life histories of some of the prominent avian secondary predators of the Antarctic.

BLUE-EYED SHAG

The Blue-eyed Shag (also known as the Imperial Shag) is a bird that figured importantly in early Antarctic explorations. The shags or cormorants are known to sailors the world over as birds that seldom stray far from land. Thus, in January 1775, when Captain James Cook discovered the island of South Georgia he recorded, ''The shags and soundings were our best pilots.'' In 1916, when Ernest Shackleton was on the final leg of his incredible escape from the Weddell Sea ice pack, descendants of those same Blue-eyed Shags alerted him to his first landfall at South Georgia after crossing more than 1,200 kilometers of Southern Ocean in a canvas-covered whaleboat.

The cormorants or shags (family Phalacrocoracidae) form a group of 27 or 28 species of large water birds that feature hooked bills, long necks, and elongated bodies and tails. They swim well underwater—diving to depths as great as 30 meters—where they pursue the fish that are important in their diets. The Blue-eyed Shag, *Phalacrocorax atriceps,* is a wide-ranging cormorant of the Southern Hemisphere that breeds on the Antarctic Peninsula, and islands of the Scotia Arc and the subantarctic. It is also a common breeder along the Argentinian and Chilean coasts of South America. Because of its habit of not flying far from land, this bird forms fairly isolated local breeding populations on the oceanic islands that it occupies. Consequently it presents a variable (polymorphic) appearance

"The shags seem to pick very exposed spots to nest—whole cliff sides. Occasionally they preen each other and sway their heads and necks back and forth in circular patterns."

and shows differences in the details of the color patterns of its head and wings across its range. The shags of the Peninsula (considered with those of the Scotia Arc to comprise the subspecies, *P. a. bransfieldensis*) are striking birds in their adult breeding plumage. The Blue-eyed Shag has a white belly and throat, with a shiny blue-black back. The long hooked bill is brownish with bright orange or yellow caruncles at its base. The brown iris of the eye contrasts with the ring of bright blue skin that surrounds it, giving this cormorant its common name.

Immature birds are brown on the back and reach adulthood only after three or four years. When full grown they reach about 72 centimeters (28 inches) in length with wingspreads of about 124 centimeters (49 inches). They breed in scattered rookeries, usually on small islands, with as many as several hundred pairs occupying the same site. The rookeries are used as roosts and gathering places the year around (shags are decidedly social), and accumulate thick deposits of guano over time. Shag guano, along with generous quantities of lichens, mosses, and seaweeds, is used by the birds to construct their nests. A great deal of thievery of nesting materials takes place within the colony—Blue-eyed Shags are genuine kleptomaniacs. The nests are truncated, cratered cones, 20 to 25 centimeters high and about 40 centimeters in diameter, with deep central depressions. By the end of the nesting season the nests are trampled down to much lower versions.

Nest building is accompanied by extended bouts of courtship that include the pair standing side by side, placing their cheeks together, and then bowing in concert. The necks are then twisted to bring the opposite cheeks together and bowing occurs once again. The movements are repeated for long intervals, broken occasionally when the male departs for a brief flight. The courtship continues even after the eggs have hatched and the young are reasonably well developed. Blue-eyed Shags use the same nesting site and choose the same mate year after year.

The first eggs are laid in late October or early November. Blue-eyed Shags differ from other nesting birds of the Antarctic in that the young are naked and incapable of temperature regulation when they hatch (altricial chicks). They are therefore exceptionally dependent on their parents for early care, and, indeed, both the mother and father are in nearly constant attendance at the nest before and after hatching. Hatchling Blue-eyed Shags vary more in body temperature and attain true homeothermy (constant body temperature) slower—nearly 2 weeks—than South Polar Skuas (1 to 2 days), Southern Giant Petrels (3 days), and Adélie penguins (1 day). Despite this the young grow rapidly and develop faster than penguins hatched at the same time. Fledging occurs in March.

Like most nesting birds of the Antarctic, shags are subject to disastrous

Hooked bill and caruncle of shag

Blue-eyed Shag on nest

weather conditions that kill large numbers of young in certain years. The precarious environment is a major factor in determining the advantage of being long-lived. Parents that lost their young in any given reproductive season have additional chances for genetic immortality as they survive from one breeding year to the next.

Unlike most antarctic birds, Blue-eyed Shags are not migratory, moving in winter only far enough north to reach open water for feeding. The

"Many nests have three chicks. When they are full size they really fill up the nests, and kind of hang over the edge— big, floppy, dark grey birds, uncoordinated and spastic looking. They seem to be constantly hungry, squawking and vibrating their heads from side to side, and reaching up to find the parent's beak. The entire head of one chick was engulfed in the parent's throat. All I could see was its neck and the adult with a bulging throat. It was a big chick too!"

food is principally fish, particularly the smaller species of ice fishes. Occasionally they capture krill and other crustaceans, but the bulk of their diet places them at the secondary predator level of the ecosystem.

Blue-eyed Shags are locally abundant around breeding rookeries or wintering areas, and concentrations of thousands of birds have been reported. Their rocky island nesting sites and shallow water feeding grounds, however, are in comparatively short supply in the Antarctic, except on the Peninsula where they thrive. Considering the immensity of the Southern Ocean, their average density is extremely low.

AMERICAN SHEATHBILL

Of all the birds of the Peninsula, the American Sheathbill *(Chionis alba)* is probably the oddest and most unexpected. In a land where the birds are tied tightly to the sea, the sheathbill is the closest thing Antarctica presents as a terrestrial form. If you picture a white, grouse-sized bird, without webbed feet; and one which struts along, bobbing its head like a pigeon, you have a good image of an American Sheathbill.

"Two American Sheathbills appeared while I was there and wandered through the colony, being chased off as much as possible by the shags who squawked and reached out as far as their necks would allow. One shag was sitting beside an egg with a hole in the top—no doubt from the sheathbills. They eventually got the entire egg."

The term "sheathbill" refers to a horny sheath that covers the nostrils. There are also fleshy protuberances at the base of the bill. Two species of sheathbills (family Chionidae) exist; with the second species, the Lesser Sheathbill, found only on the subantarctic islands of the Indian Ocean. Some adults of the American Sheathbill are nonmigratory and live on the Peninsula through the Antarctic winter, but many adults and most adolescents fly north in the winter as far as the Falkland Islands and the South American coastline.

Much of the known life history of the American Sheathbill is fascinating in its contrast with that of other Antarctic birds. This one is not a conspicuous colonial nester, being instead a cryptic, solitary breeder whose nesting sites are often within rookeries of penguins or shags. Near Palmer Station sheathbills are known to burrow a meter or two through accumulations of shag guano to create a nesting chamber. Eggs (usually two or three) are laid toward the middle of December, and hatch about four weeks later. The young are soon active, running about the nest site; and have feathered out by the middle of February.

Sheathbills are most frequently found on dry land, but they also are known to eat mollusks, crustaceans, and algae in the shallow waters at the shoreline. They avoid deep water, although despite their webless feet they can swim well. Their flight is pigeon-like, but strong enough to carry them over the Drake Strait to South America. Many reports exist of them at sea three or four hundred kilometers from the nearest land. Sheathbills fairly frequently will land on the decks or rigging of ships far at sea in the high southern latitudes. This is where Charles Darwin encountered them while on board H.M.S. Beagle.

American Sheathbills are bold, garrulous, and unafraid of virtually anything except skuas. In their feeding habits they are best described as scavengers, although they are noted predators of eggs and chicks in the penguin and shag rookeries that they roam. They quickly locate the carcasses of dead birds, seals, and other sorts of carrion. Sheathbills burrow under incubating (and unsuspecting) Blue-eyed Shags to steal eggs. Although they themselves lay two to three eggs, only one of these usually survives. This leads to the suspicion that they may cannibalize their own eggs or chicks; just as skuas are known to do. In addition to carrion, eggs, and chicks, sheathbills eat food regurgitated by penguins, penguin droppings, and the placentae, umbilical cords, and blood discharged by birthing elephant seals. This bird lives at the extreme end of the Antarctic food web where the percentage of the sun's energy that originally rained on the Antarctic is as but the glimmer of a distant star.

Except for local aggregations of one hundred or so sheathbills in the northern wintering range, the population densities of this bird are quite low throughout the Antarctic Peninsula and the Scotia Arc. A census on South Georgia many years ago found only one sheathbill per 3 kilometers of shoreline. American Sheathbills that spend the winter on the Peninsula often gather in the vicinity of research stations to feed on handouts and refuse. Much is yet to be learned about their normal winter diet. They are out and about in all but the most extreme weather, and have been reported normally active at temperatures below -40° C. Altogether the American Sheathbill is an unusual bird by any standard—even that of Antarctica!

SOUTHERN GIANT PETREL

The Southern Giant Petrel (or southern Giant Fulmar), *Macronectes giganteus,* is the largest of the petrels (family Procellariidae) and superficially resembles its albatross relatives, with which it might be confused by neophyte bird watchers. This bird, however, lacks much of the grace aloft of the albatrosses, as well as the planktivorous food habits of the smaller petrels. In *M. giganteus* we find a bird that manages, vulture-like, to offend the sensibilities of those ornithologists and seafarers who are familiar with it. The first hint of this dislike comes from its unofficial common names that include "stinker," "stink-pot," "glutton," and "sea vulture." Moreover, the Southern Giant Petrel is generally described as "ungainly and uncouth," and has been likened to other birds as sharks are to fish. Its food habits, defensive behavior, behavioral attitudes on land and in the air, and general rapaciousness all earn this reputation for it. If nothing else this bird's distinctive adaptations demonstrate the difficulties inherent in carving out an existence for a top antarctic carnivore.

This petrel is one of the larger sea birds, reaching almost a meter in length with wingspreads that often exceed two meters. Although the sexes are alike, males are larger than females. Its plumage is singularly undistinguished. Young birds start off life with a black plumage that in time

"I saw a white phase giant petrel—almost white with just a few dark patches. They cross land with great difficulty, running awkwardly, flapping their wings at the same time. They mostly sit or stand or fly."

wears and weathers to a dirty brownish color. Gradually white feathers intrude, so that with maturity (that comes only after seven to eleven years) the head, throat, and leading edges of the wings are white merging to a dark, speckled coloration farther aft. The large, heavy, and strong bill with its prominent tubular nose is greenish at the tip, separating it from the closely related Northern Giant Petrel *(M. halli)* with which it mingles on the subantarctic islands in the northern part of its breeding range.

Like the Blue-eyed Shag, *M. giganteus* is a polymorphic species with less than 10 percent of the population existing in a white color phase. The basic white is broken only by dark mottlings scattered irregularly over the body and wings. In the breeding population near Anvers Island only one out of fifteen to twenty birds is of the white phase.

The Southern Giant Petrel is another bird of comparative great longevity. After fledging, the young birds disperse downwind from its breeding colonies on the Peninsula. Birds banded in South Georgia have been recovered as far as 16,000 kilometers away in Western Australia and New Zealand only five weeks after fledging. Immature giant petrels roam the Southern Hemisphere, penetrating as far north as 10° South latitude off the western coast of South America. At about age six or seven the young birds appear in the nesting colonies and attempt to nest. They have little success until their tenth or eleventh year. Birds banded as adults have been recovered fifteen years later, still nesting in the same colonies, so this bird can live at least 25 years, and probably much longer. Some estimates range as high as 50 years.

This species nests in open colonies on relatively flat land. The nest is built of just about anything available, such as shells, moss, grass, feathers, bones, and stones. The colonies range from a few birds to a thousand or more. The nest usually contains but a single egg. The awkwardness of

Giant petrel and chick

these birds allows skuas to abscond with eggs left uncovered, and in certain colonies predation by skuas is high. Skuas constitute practically the only natural enemy of the Southern Giant Petrel.

After an incubation period of up to 48 days the chicks hatch and rapidly become homeothermic. After more than 100 days of growth the young are no longer fed and wander about for as long as a month before they begin flying, finally dispersing around the first of May.

There are precautions to be taken when visiting a colony of Southern Giant Petrels. Adults incubating eggs or newly hatched young have been known to lash out with razor sharp bills that can slash flesh and cut through heavy shoe leather. Continued harassment can provoke the bird to open its mouth and spit a bolus of foul, molasses-like stomach oil toward the offender. Even day-old chicks have this defensive ability. The clinging, musky stomach oil is characteristic of all procellariiforms but the diving petrel. Usually a clear orange or red in color and found in volumes sometimes exceeding 200 milliliters, the oil is secreted from glandular cells in the proventricular stomach. It has appreciable quantities of vitamins A and D (like cod liver oil), and in its hydrocarbon composition is similar to the spermaceti found in sperm whales. Suggestions about its use include the defensive purposes described above and as a food to feed young chicks.

Giant petrels taking to the air

The aerial artistry of Southern Giant Petrels is not as graceful as many sea birds. They fly with a labored, stiff beating of wings, punctuated at intervals by brief glides. In calm weather they rise with difficulty from the water surface, but in stronger winds they handle themselves adroitly, scaling in the breeze. Fortunately for them the Southern Ocean is not a place known for its calms. The Southern Giant Petrel is the only bird of its tribe to feed on land, where it eats carrion and even attacks other birds and small mammals (rats and rabbits have been introduced onto some of the subantarctic islands) that may be available to it. On land it maintains its reputation as an awkward, clumsy bird.

In its normal food habits the Southern Giant Petrel is foremost a scavenger, feeding on dead whales, seals, and birds of all kinds. Its keen eyesight and sense of smell allow it to congregate at the sites of such carrion. Vulture-like, it is said to eat a hole through the skin of a whale or seal, then consume the softer tissue underneath. Seasonally these birds gather to gorge on Weddell seal placentae in the pack ice in early spring (adults are year-round residents of the Peninsula and edges of the pack ice), as well as to prey on the eggs and chicks at penguin rookeries. Indeed, any bird or other creature smaller than itself is subject to attack from Southern Giant Petrels. Reports exist of their catching other petrels on the wing, killing and eating them. Unsubstantiated stories from explorers and sea-farers attribute attacks by Southern Giant Petrels on sailors lost over-board. Stomach contents recorded from them include squid and crusta-ceans, young shags, stones, seaweed, rope, and bird remains. No wonder they are sometimes depicted as sharks of the air!

Despite this reputation for scavenging, Southern Giant Petrels can drop down a trophic level and feed on a substantial amount of krill—as much as 40 percent of its diet by some estimates. What we have here is a bird that carves a living out of an environment, which, at its trophic level, is not rich. The pie is not large, so the Southern Giant Petrel has to take a big slice. This means getting food from a diversity of sources, and using techniques that don't always strike the human observer as being fair. But "fairness" in our terms would spell the end of this bird. We should ad-mire it instead for its persistence in the face of a rigorous, even hostile, environment.

SOUTH POLAR SKUA

If the Southern Giant Petrel is the vulture of Antarctica, the South Polar Skua *(Catharacta maccormicki)* is surely its eagle. Although skuas are the equals of giant petrels as scavengers and opportunistic feeders—and even exceed them as fierce predators—they raise their mayhem with consid-erably more grace and style. Their defense of breeding territory, piratical hunting techniques, and fearlessness of humans help them win the grudging admiration of all who observe them closely.

South Polar Skua

The six or seven species of skuas and jaegers found worldwide comprise the family Stercorariidae, which are gull-like but with a distinctive bill structure. The skuas are heavy, broad-winged, brownish birds with a pair of slightly projecting feathers in the center of the tail. These are wide-ranging sea birds of high latitudes, and have a taxonomy that is still not agreed upon by experts. The Southern Polar Skua is clearly the most southern-living bird in the world—having been repeatedly recorded hundreds of kilometers inland from the Ross Ice Shelf, toward the South Pole. It nests along the entire coast of the Antarctic continent, including the Peninsula. In this latter location its breeding range overlaps with that of another skua, which was originally thought to be a southern form of the Great Skua, *C. skua*, that breeds in the North Atlantic. Recent studies have proposed that this second Peninsular skua is a subspecies *(lonnbergi)* of the Antarctic Skua, *C. antarctica*, (also known as the Brown Skua) that breeds on islands of the antarctic and subantarctic as well as on the Peninsula. Of biological interest is the fact that both of these skuas are polymorphic, having distinctly different grey, brown, and intermediate color phases. Even though the South Polar Skua is a smaller bird, and nests somewhat later in the year, it freely hybridizes with the Antarctic Skua along the northwestern rim of the Peninsula. Numbers of hybrid chicks have been banded, and their reproductive fate is being watched with interest by antarctic ornithologists.

The South Polar Skua appears to be one of the world's widest ranging birds. Unlike the Blue-eyed Shag, the American Sheathbill, and the Southern Giant Petrel that remain in antarctic residence during the winter, these skuas depart, usually during the month of May. Sightings and band returns suggest that they regularly move in a clockwise gyre throughout the great oceans (Indian, Pacific, Atlantic), penetrating into far northern latitudes before returning to Antarctica the following October. Banded birds have been found in northern Japan, the Aleutians, and California in the North Pacific; northeastern America to Newfoundland, Greenland, and Great Britain in the North Atlantic.

On the breeding grounds South Polar Skuas are social birds, with nests (simple scrapes in the ground) as close together as fifteen or twenty meters apart. Although their repertoire of displays is more limited than that of most gulls, they have a noteworthy and characteristic threat display that features the breast puffed out with the wings held up high over the back like a symbolic Viking eagle. People intruding into their nesting colonies are greeted with shrieks, threatening dives, and buffeting with wings and feet. When undisturbed, South Polar Skuas communicate with each other with soft duck-like quacking sounds.

On the Antarctic Peninsula South Polar Skuas arrive for the austral summer by the end of October. Breeding territories are claimed with the

males staking out the same general location year after year. The birds begin to breed as five year-olds with but a single egg produced. Banded birds have survived for at least 20 years, and, as they age, the clutch size increases to two. Adult mortality is low, with only 6 percent of the mature population succumbing in a normal year. Outbreaks of fowl cholera have decimated certain colonies in past years, and chick mortality can be high if unseasonable blizzards occur early in the rearing period.

The eggs are laid beginning in late November, and hatching commences about one month later. South Polar Skuas feed their young chicks first on krill, and later switch over to fish, penguin eggs and chicks, and whatever else they can provide. The food is deposited on the ground near the nest and the chicks must find it themselves. This apparently encourages the chicks to be active and resourceful. Chicks that roam too far from the nest tempt neighboring adult skuas. Cannibalism is a source of chick mortality. After a period of rapid growth the young fledge between

Young skua chick

An older skua chick

the middle of February and the middle of March. All of the birds depart for their oceanic wintering grounds during the month of May.

On the Peninsula the South Polar Skuas outnumber the larger, darker Antarctic Skuas by about eight to one. The latter birds nest about two weeks earlier, and some studies have shown they rely on penguin eggs and chicks to a greater degree than do the South Polar Skuas, which they tend to exclude from the penguin colonies.

The food habits of the South Polar Skuas have been especially well studied at breeding colonies near Cape Bird on Ross Island in the Ross Sea. Here 250 pairs of skuas nest in proximity to a colony of about 24,000 pairs of Adélie penguins. Contrary to popular notions about their reliance on penguins, these skuas feed mainly at sea, returning with fish (Antarctic silverfish), crustacea, and squid. Penguin eggs and chicks are taken, however, and by one estimate these birds consumed over 23,000 penguin eggs in a single season. After the penguin chicks hatch they are preyed on by skuas for quite a period of time. Skuas are powerful birds and are capable of lifting and carrying their own weight (more than a kilogram) in prey.

In addition to feeding on penguins, skuas eat krill and fish spilled by the adult penguins as they feed their young. They are also notorious pi-

Skua in penguin colony

rates and will harass shags, petrels, and gulls until they disgorge their prey. The skuas have an uncanny knack for recognizing which sea birds have recently fed, and which therefore are worthy of harassment. If their aerobatically sophisticated chases do not cause the prey bird to regurgitate, they may seize a petrel or shag by a wing and topple it into the water.

Small petrels like Wilson's Storm Petrel or Dove Prions leave and enter their nesting burrows at night because of the danger of skua predation. One of these birds afoot in the daytime is no match for the skua. Other sources of food include carrion of all sorts. Skuas quickly learn to visit manned research stations throughout the Antarctic to receive handouts. Their individuality and élan make some of them favored "pets" of personnel at these sites.

It is as harassers of penguins that South Polar Skuas display the resourcefulness that makes them successful in spite of the energetic hardships facing top predators. Skuas adopt several strategies to overcome the defenses of the much larger brooding adult penguins. Generally it is the peripheral nests of a group of nesting penguins that are attacked successfully (penguins nesting along the colony's margin are usually the youngest and least experienced of the breeding birds). Attacks can be made by skuas walking on the ground, or by flying birds from above. In the former case the walking skua stabs under the brooding penguin with its strong hooked beak. These attacks can be made from the front, side, or rear of the penguin. When attacking from behind the skuas pull the penguin by its tail, attempting to drag it off the nest. Two skuas occasionally cooperate in this venture, one pulling from behind, the other attacking from the front or side. While strolling through the penguin colony a skua sometimes takes the strategy of suddenly jumping at a defended nest from a distance of about five meters hitting the brooding penguin with its feet and trying to topple it long enough to grab the egg or newly hatched chick. When attacking penguin colonies from the air (two or more skuas are always in attendance at the rookeries, making low-level cruising flights) they are alert for untended nests, eggs, or moribund chicks. They may stoop low over the peripheral nests, distracting the defending penguins. In the most active attacks they will fly directly into the adult penguin, buffeting it and attempting to ram it away from the nest. Once an egg or young chick is exposed, the skua picks it up in its mouth in a flash, and departs for its own nest site with scarcely a pause.

When the growing penguin chicks are in the creche stage, skuas sometimes fly at the feeding penguins, causing food to be spilled. Chicks which wander away from the main colony will be attacked from the ground or the air. The skuas grab the chick by the head and attempt to carry it away.

South Polar Skuas are indeed interesting predators. Their strength and agility in the air always invite comparison with the hawks and falcons of lower latitudes. Yet the South Polar Skuas in their communal nesting habits, in their displays and vocal communication, and in their cooperative hunting techniques are unlike most other predatory birds. This combination of sociality and predatory behavior approaches that seen in mammals like lions, wolves, Killer whales, and humans. Perhaps this is why we find skuas to be such fascinating animals.

EMPEROR PENGUIN

Although the *Pygoscelis* penguins are fascinating and marvelous birds, their fame is overshadowed by an even more spectacular one, the Emperor penguin (*Aptenodytes forsteri*). This, the world's largest penguin and diving bird—adults average about 32.5 kilograms—is also the only bird to lay eggs and brood its young on the Antarctic Continent during the dead of winter. These aspects of its biology, plus the fact that it can dive deeper in the water than any other bird, make its natural history particularly interesting.

Emperor penguins enjoy a more southern distribution than does any other penguin. With the exception of occasional breeding colonies on the Peninsula (the best known being in the Dion Islands), these birds breed on fast ice that is firmly attached to East Antarctica and its nearby islands. Just a handful of sightings of Emperors—all at sea—have been made as far north as the Scotia Arc.

These penguins are large. Adult Emperors can extend their bodies to heights of about 1.3 meters with necks outstretched. The more usual pose, however, is with the head tucked back against the shoulders, with the birds standing only 85 centimeters or so. With their long, slightly curved bills, and sleek dense plumage (the densest and longest feathers of all penguins) Emperors are impressive animals. They present a white belly with a handsome slate-gray back that is sharply set off from the belly by a distinctive black edging passing under the wings. The head is black with conspicuous light patches below and behind the eyes. As striking as these birds are, they fall short of the colorful standard set by their near relative, the King penguin that breeds only on islands of the subantarctic. This latter bird, although somewhat smaller than the Emperor, carries a golden comma-shaped mark on either side of its head, and has a more yellowish appearance with its ocher breast feathers.

"We saw two Emperor Penguins the other day on a floe right next to the ship—sleek, but awkward looking birds."

Emperor penguins are social both in and out of the water. They swim in groups, and dive together to search for food, sometimes diving beneath sea ice through holes that may be many kilometers from open water. They swim with their heads drawn back against their bodies and their feet held straight out behind, soles upward. The power for swimming comes from their long trim wings. Measurements of swimming speeds made on birds released from one hole in the ice and required to swim to a distant one show them to move at velocities that range between 5.4 and 9.6 kilometers per hour. To do this the penguins beat their wings between 24 and 49 times per minute and covered as much as 4.8 meters with one beat of the wings.

Groups of normally foraging Emperors spend between 2.5 and 9 minutes underwater per dive. One experimental penguin stayed underwater for at least 18 minutes. Emperors make dives of much greater duration than do Chinstraps and (presumably) other penguins. A depth recorder fastened to the back of a feeding Emperor penguin showed it to dive as deep as 265 meters (870 feet). Because this bird was encumbered with a heavy (700 gram) and bulky backpack containing the recording device, unfettered Emperors can probably dive more deeply. Even so, 265 meters is an impressive depth. This puts the bird under a pressure of more than 26 atmospheres. Unlike seals, which exhale before dives and collapse their lungs at depth, the penguins deeply inhale just before going down.

Bird lungs are fundamentally different in structure from those of mammals. They are proportionately smaller, and the gases flow through the lungs in a way that is more efficient than the tidal means by which mammals ventilate their lungs. Other diving birds (ducks) are known to slow down the heart rate substantially when underwater, and to direct the reduced blood flow only to those organs (e.g., the brain) which need oxygen the most. It is conceivable that Emperor penguins are able to dive so deeply and so rapidly because they minimize the amount of gas in the lungs which could be dissolved in the blood under pressure. The deep dives of these penguins allow them to feed at depths where their favorite food—squid—is found, even in the dark of the winter. Unlike most other

Antarctic birds they do not have to retreat to the north to find food at this most inhospitable time of year.

Because Emperors are so large, their chicks have a lot of growing to do in order to reach fledging size. Ideally fledging should correspond with the summertime period of great food abundance and availability. To achieve that goal Emperor penguins get an early start on reproduction. Indeed, they start so early in the season that their eggs are laid and incubated in the dead of the antarctic winter—surely one of the most unusual reproductive modes in the animal kingdom!

This odyssey begins in late March when solid new sea ice forms. The Emperors trek in single file across this ice to the location of traditional rookeries that typically are located on strong fast ice that will not break up until late in the next summer. The journey can carry the penguins as much as 120 kilometers before they reach their destination. After courting and mating in late April, the female lays a single egg that weighs 450 grams (one pound) and engages in ritual displays that include special vocalizations. It will be by sound that the male and female identify each other later in the season. Having lost nearly 25 percent of her original body weight from the migration, courtship, and egg laying, the female leaves the egg in the care of her mate and departs across the sea ice to feed in the open water beyond. Because a month's new ice has formed, this return trek is substantially longer than was the original one on the way in to the rookery.

The assembled males face a two-month incubation period through May, June, and into July—the deepest part of the winter. During this period intense katabatic and hurricane force storm winds rage around them while temperatures sometimes drop to between -25° and -49° C. Unlike all other penguins, excepting its near relative the King, the Emperor incubates its egg not in a nest on the ground, but balanced on top of its two feet and tucked under a large loose fold of abdominal skin that keeps the egg warm. This behavior restricts the penguins to careful shuffling movements when walking, and precludes them from choosing anything but flat surfaces on which to locate their rookeries.

Attempting to find relief from the rigors of this extreme environment the male penguins crowd together to form "huddles" that protect those birds in the core of the huddle from the wind and cold. The unfortunate Emperors on the windward edge of the mass gradually work their way (or are blown by the wind) along the sides of the huddle to finally reach shelter in the lee. These birds are in turn enclosed by new arrivals from the front, and pressed through the huddle to once again emerge to windward. In this way the huddle gradually circulates and moves downwind, exposing each penguin to its equal share of the hostile environment.

In the ideal Emperor world the female, having fed in the distant open

Huddle of Emperor Penguins

water and regained her fat stores, completes the return trek to the rookery in time to relieve her mate just about the time the egg hatches. But this ideal world is fraught with dangers. The female must navigate across ice that did not exist on her outbound trip. Where the original trek in may have been 100 kilometers, the next return may be more than 300 kilometers—and this through continual night and shrieking gales. The navigational system that allows her to make such a jaunt, and find a huddle of perhaps a few thousand males, covering only a few hundred square meters—truly a needle in a haystack—can only be marveled at.

The male, although physically inactive, has continued to burn his energy reserves as he keeps himself and his tended egg warm. As hatching time approaches his fuel supply runs dangerously low. Studies on incubating Emperors indicate a critical body weight of about 22 kilograms. If the male's weight drops to this point he will usually abandon the egg or newly hatched chick and begin his walk to the open sea, in effect making the assumption that his mate will not return, and that their chick is doomed. The least he can salvage in his own life and the chance to breed again in another year.

With surprising frequency, however, this unfortunate scenario does not occur. The female arrives, identifies her mate by his voice, and begins her stint of brooding. If the chick hatches under the male's care he can feed it milky secretions from his esophagus. When the female arrives she carries undigested krill, fish, and squid in her stomach that she regurgitates to feed the chick. Having been relieved (this is now mid-July) the male makes his way to open water where he feeds to rebuild his lost body reserves. After four weeks of this he starts back to take another incubation turn.

Throughout the late winter and spring these alternating periods of brooding by the parents continue, with the over-ice treks becoming progressively shorter as the sea ice breaks up. By November a chick can get as much as 4 kilograms of food in a few hours from its parents. A chick that has lost its parents is usually doomed, for the adults nearly always feed just their own offspring, which they identity by its sounds.

By December the chicks are ready to molt to their juvenile plumage. This process must be completed before the rotting sea ice under their feet breaks up. Chicks still in down have been seen floating away on deteriorating floes, probably to succumb when dumped into the icy water while lacking the waterproofing and insulation of the permanent plumage. At fledging (which occurs about 170 days from hatching) the chicks are only 60 percent of the adult body weight (most birds are 100 percent, or more, at this stage). This appears to be a compromise between the shortness of the summer season, and the necessity to get the young penguins in the water and feeding during the annual bloom of plankton. Overall the mortality of chicks in this perilous environment is high. In exceptionally bad years as many as 90 percent of all the chicks in some rookeries have failed to survive. The annual mortality of adults, however, is much lower, averaging only about 5 percent of the population. Clearly, these birds must live a long while in order to expect the kind of reproductive success necessary for their genetic survival.

OTHER AVIAN SECONDARY PREDATORS

In addition to the Blue-eyed Shag, American Sheathbill, Southern Giant Petrel, skuas, and Emperor penguins there are a few other Antarctic birds that derive some or most of their living from eating other predators. Prominent among them is the Kelp Gull *(Larus dominicanus)*, also known as the Dominican Gull or Southern Black-backed Gull. This bird breeds

Kelp Gull

Regurgitated limpet shells

on the Peninsula to about 65° South, as well as through much of the Southern Hemisphere south of the Tropic of Capricorn. In its predatory behavior and scavenging food habits it greatly resembles the anatomically similar Great and Lesser Black-backed Gulls of the Northern Hemisphere. On the Peninsula breeding Kelp gulls feed to a great degree on the limpet, *Nacella concinna*, which they find in the intertidal zone. Having digested the soft parts of a meal of limpets, the gull regurgitates a nicely nested pack of the limpet shells that end up littering the beach areas that these gulls frequent.

A number of the larger petrels, even though deriving much of their energy from krill at the primary carnivore trophic level, will not hesitate to scavenge whale and seal offal if the opportunity presents itself. Conspicuous among these part-time secondary carnivores are the Antarctic Fulmar *(Fulmarus glacialoides)* and the Cape Petrel *(Daption capense)*. Among the three species of penguins commonly breeding on the Peninsula, the Gentoo *(Pygoscelis papua)* is most likely to take some fish and squid in its diet in addition to krill. This is reflected both in the smaller population size of this species and in the color of its guano, which is whiter than the orange, krill-derived wastes of the Adélie and Chinstrap penguins.

When the major predatory birds of Antarctica are reviewed, several aspects of their biology stand out. First is the relatively small size of their populations when compared to the immensity of the resources of the Southern Ocean. This, of course, is a consequence of the inefficiency of the conversion of energy from one trophic level to another. Second is the range of food types sought and accepted by these predators. All of these birds can and will eat virtually anything. Third, aside from the American Sheathbill, all of these birds live in a social context. This could reflect a shortage of nesting sites, but communication seems to play an important role in their lives, and some, like the South Polar Skua, engage in a rudimentary form of cooperative hunting. Fourth, all of these species grow

Gentoo penguin

to ripe old ages once they survive the perils of the first few years of life (this is common to sea birds the world over). In some years chick mortality can be high, but in every year the newly fledged birds face a hazardous existence as they launch themselves into the hostility of the autumnal Southern Ocean. The extended breeding lifetimes of these birds are requisite to provide for the eventual replacement of themselves in the population. Finally, it is striking that most of these species display a lot of morphological variation or polymorphism—more apparently than that seen in the larger populations of avian primary predators. It would be interesting to know if this variability in color pattern and appearance is underlain by similar diversities in physiology and behavior. Perhaps in the extremity of the antarctic environment small populations survive best when a wide range of phenotypes is available to meet every possible disaster.

It will be worthwhile to compare the patterns of life presented by these birds with those seen in the mammalian secondary predators of the waters surrounding the Antarctic Peninsula. We will do that in the next chapter.

ADDITIONAL READING

Bernstein, N. P. and S. J. Maxson. 1984. Sexually distinct daily activity patterns of Blue-eyed Shags in Antarctica. *The Condor,* 86:151–156.

Croxall, J. P. 1984. Sea birds. In R. M. Laws, ed., *Antarctic Ecology,* 2:533–616. London and Orlando, Fl.: Academic Press.

Fraser, W. R. and D. G. Ainley. 1986. Ice edges and seabird occurrence in Antarctica. *Bioscience,* 36(4):258–263.

Hemmings, A. D. 1984. Aspects of the breeding biology of McCormick's *Skua catharacta maccormicki* at Signy Island, South Orkney Islands. *British Antarctic Survey Bulletin*, no. 65, pp 65–79.

Hunter, S. 1985. The role of Giant Petrels in the Southern Ocean ecosystem. In W. R. Siegfried, P. R. Condy, and R. M. Laws, eds., *Antarctic Nutrient Cycles and Foods Webs*, pp. 534–542. Berlin: Springer Verlag.

Kooyman, G. L., C. M. Drabek, R. Elsner, and W. B. Campbell. 1971. Diving behavior in the emperor penguin *Aptenodytes forsteri*. *Auk*, 88:775–795.

Kooyman, G. L., R. W. Davis, J. P. Croxall, and D. P. Costa. 1982. Diving depths and energy requirements of king penguins. *Science*, 217:726–727.

Laws, R. M. 1977. The significance of vertebrates in the Antarctic marine ecosystem. In G. A. Llano, ed., *Adaptations Within Antarctic Ecosystems*, pp. 411–438. Washington, D. C.: Smithsonian Institution.

Laws, R. M. 1985. The ecology of the Southern Ocean. *American Scientist*, 73:26–40.

LeMaho, Y. 1977. The emperor penguin: A strategy to live and breed in the cold. *American Scientist*, 65:680–693.

Maxson, S. J. and N. P. Bernstein. 1982. Kleptoparasitism by South Polar Skuas on Blue-eyed Shags in Antarctica. *Wilson Bulletin*, 94(3):269–281.

Maxson, S. J. and N. P. Bernstein. 1984. Breeding season time budgets of the Southern Black-backed Gull in Antarctica. *The Condor*, 86:401–409.

Müller-Schwarze, D. 1984. *The Behavior of Penguins Adapted to Ice and Tropics*. Albany: State University of New York Press.

Murphy, R. C. 1936. *Oceanic Birds of South America*. 2 vols. New York: American Museum of Natural History.

Offredo, C., V. Ridoux and M. R. Clarke. 1985. Cephalopods in the diets of Emperor and Adélie penguins in Adélies Land, Antarctica. *Marine Biology,* 86:199–202.

Parmelee, D. F., W. R. Fraser, and D. R. Neilson. 1977. Birds of the Palmer Station area. *Antarctic Journal of the United States,* 12:15–21.

Parmelee, D. F. and C. C. Rimmer. 1985. Ornithological observations at Brabant Island, Antarctica. *British Antarctic Survey Bulletin,* no. 67, pp. 7–12.

Pietz, P. J. 1985. Long call displays of sympatric South Polar and Brown Skuas. *The Condor,* 87:316–326.

Stonehouse, B. 1970. Adaptation in polar and subpolar penguins (Spheniscidae). In M. W. Holdgate, ed., *Antarctic Ecology,* 1:526–541. New York: Academic Press.

Stonehouse, B., ed. 1975. *The Biology of Penguins.* London: Macmillan.

Voous, K. H. 1968. Antarctic birds. In J. van Mieghem and P. van Oye, eds., *Biogeography and Ecology in Antarctica,* pp. 649–690. The Hague: Junk.

Watson, G. E. 1975. *Birds of the Antarctic and Subantarctic.* Washington, D. C.: American Geophysical Union.

Watson, G. E., J. P. Angle, P. C. Harper, M. A. Bridge, R. P. Schlatter, W. L. N. Tickell, J. C. Boyd, and M. M. Boyd. 1971. Birds of the Antarctic and Subantarctic. *Antarctic Map Folio Series,* folio 14. New York: American Geographical Society.

CHAPTER 9

SECONDARY CARNIVORES— MAMMALS

The populations of mammals that fuel their lives with energy gotten from prey such as fish or penguins are small when compared to those that live on krill. The cold conditions of the Antarctic inflate the metabolic costs of doing business here. Although the direct thermoregulatory costs are not exorbitant (because of good insulation), it still requires extra energy to build that insulation, so proportionally less of what is eaten can be directed to growth and reproduction. Therefore, the efficiency of energy transfer from the preceding trophic level to this one is low. Despite the enormous stocks of krill, there is not much food to support the higher trophic levels.

Antarctic mammals, like their peers in warmer climes, resort to several behavioral and physiological adaptations to survive in a food-limited world. The special problems presented by the Southern Ocean, however, require extreme adaptations—often beyond those seen elsewhere. In the extremity of specializations for finding food atop the trophic pyramid, and for reproducing in this deep, cold, ice-laden sea, we find the quintessence of life in and around the Antarctic Peninsula. These qualities are found in toothed whales (such as the Killer and Sperm whales), the deep-diving seals (southern elephant seal, Weddell seal, Ross seal), and in the broadly adapted leopard seal. A closer look at the biology of these animals is in order.

WHALES

The toothed whales of the cetacean suborder Odontocetes show a different pattern of life than that of the baleen whales discussed in chapter 7. They are smaller on average, most have teeth (either functional or rudimentary), and all lack the baleen plates. Moreover, the odontocetes are especially social, often traveling and feeding in well-defined groups; and most are sexually dimorphic, with males larger than females. Many species have vocal repertoires of clicks and whistles that are used in communication, echolocation, and, possibly, feeding. Most important to our purposes, however, is the fact that odontocetes are fish and squid eaters; although some, like the Killer whale, have more ambitious appetites. Most of the squid and fish in the Southern Ocean eschew the cold rough surface waters and are found deep, either in the mesopelagic zone or near the depressed continental shelf bottom. The capture of these prey demands diving feats not seen in the mysticete whales.

The waters that bathe the Peninsula are home to several species of odontocetes. Our knowledge of them is dominated by the most conspicuous—Sperm whales and Killer whales. But smaller toothed whales also swim here, including Arnoux's Beaked whale *(Berardius arnuxii)* and the Southern Bottlenose whale *(Hyperooden planifrons)*, both of which penetrate south to the ice edge. These animals grow to intermediate size (7.5 to 10 meters), and—because they are not hunted—are poorly known. Arnoux's Beaked whale is a deep diver with squid dominating its diet. The Southern Bottlenose whale is unique among odontocetes in that a sizeable fraction of its food (as much as 50 percent) is made up of krill.

Other, still smaller, toothed whales known from near the Peninsula include Pilot whales *(Globicephala melaena)* and the Hourglass dolphin *(Lagenorhynchus cruciger)*, which is the only small dolphin to penetrate this far south. The biology of these whales is known only from infrequent examinations of stranded individuals, but fish and squid bulk large in their diets.

Sperm Whale

Sperm whales *(Physeter catadon)* were a significant presence in the Southern Ocean when whalers first started working these waters. Like the Right whales, Sperm whales do not sink after being killed by harpooning and

Pod of Sperm whales

were therefore a traditional target of eighteenth- and nineteenth-century whale fleets. The ancestral stock of Sperm whales in the Southern Ocean may have been as high as 170,000 individuals. Today that number is reduced to about 71,000. These figures are for male Sperm whales only, for it is a curious feature of the biology of this species that only mature, bull Sperm whales penetrate south to antarctic seas.

Cow Sperm whales, which grow only to about 11.6 meters, their calves, and immature whales of both sexes are denizens of temperate and subtropical oceans, where they form cohesive social groupings, sometimes accompanied by a "harem master" bull. The bulls grow considerably larger (maximum length, 18.3 meters; weights in excess of 60 or 70 tons), but reach such lengths only after a substantial period of social maturation that follows by several years a prolonged period of sexual maturation. A bull Sperm whale does not become reproductively competent until it is almost 30 years old. Like Antarctic fur seals, Sperm whales are strongly polygynous, with a single bull serving a harem of from ten to forty cows.

Compared to most mysticete whales, the cow Sperm whale has a long reproductive cycle. The 14.5 month gestation period is followed by two years of lactation, and then several months of rest and recuperation before another ovulation occurs. A cow thus mates only once every four or five years. When this low fecundity is combined with a high degree of polygyny the result is lots of males have lots of time on their flippers! These are the whales that migrate to the Southern Ocean as socially aloof individuals and feed on the rich stocks of squid found there.

Lower jaw and teeth of Sperm whale

An adult bull Sperm whale impresses any observer with its massive bulbous forehead that dwarfs the narrow, undershot lower jaw. This jaw has the two mandibles fused for as much as two thirds of its length and is studded with 20 to 25 teeth on each side that are large, conical in shape, and were favored by whalers for their scrimshaw art. With the lower jaw closed, the teeth fit into sockets or depressions in the soft gums of the upper jaws. Although upper jaw teeth exist in Sperm whales, they are rudimentary and do not normally erupt.

The teeth of Sperm whales are not needed for feeding. Immature and cow whales do not have functional teeth; and squid or fish are eaten whole, unmarked by teeth even in the bulls. Moreover, bulls with severely damaged jaws are not uncommon, yet they nearly always seem in good shape with normal amounts of food in their stomachs. Because tooth wear becomes evident with age (the enamel is quickly worn away, hence the utility of these teeth for scrimshaw), the teeth must be used, even if not for feeding. Their exact purpose is unknown, but fighting for access to females is a good guess.

The stomach contents of antarctic Sperm whales are always dominated by squid, especially those of the genera *Architeuthis, Onychoteuthis,* and *Moroteuthis.* Some of them are bathypelagic forms and tell us that Sperm whales can dive at least 1,200 meters; well below the point (100 to 200 meters) to which visible light penetrates. How the whales are able to find their prey in these dark depths is a question, although most species of deep-sea squid have bioluminescent organs and may be visible to the whales.

Sperm whales have a well-developed array of vocalizations that include clicks and trains of clicks consistent with sounds used by other odontocetes, such as the Bottlenose dolphin *(Tursiops truncatus),* for echolocation. Although no one knows for sure if Sperm whales use these clicks for echolocation, it would be a surprise if they did not. Even so, the detection of a soft-bodied animal like a squid, which does not reflect sound well, might be difficult for even a sophisticated sonar system.

Other problems posed by a squid diet include the high maneuverability and speed of squid when compared to that of Sperm whales. Calculations show that the acceleration required for a whale to catch a speeding squid may be more metabolically expensive than the energy return from eating it.

Sperm whales may be able to focus and intensify their vocal energy to the point that these sounds can immobilize prey such as a squid or fish. A key to this is thought to be the huge forehead that contains the "junk" and spermaceti organ so prized by whalers. Smaller odontocetes have bulbous foreheads ("melons") that focus beams of sounds for echolocation. The forehead of Sperm whales probably does this as well.

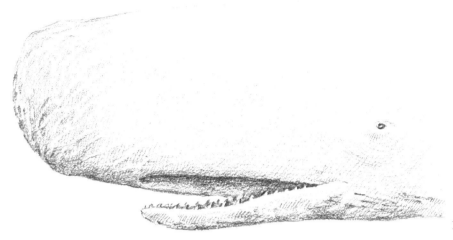

Head of bull Sperm whale

Above and in front of the Sperm whale skull lies the junk that contains wedges of connective tissue separating lenses of spermaceti (a clear, liquid wax), that terminate just above the tip of the snout where sound waves emerge from the animal's head. Above the junk rests the large spermaceti organ that is encased with thick muscle and connective tissue (the "case"). The left (functional) nostril angles up through this arrangement from the internal nares to emerge in a blow hole quite near the tip of the snout. The smaller right nasal passage ends in a series of air sacs or diverticuli that ramify below and at both ends of the spermaceti organ.

This giant complex system of muscle, connective tissue, air sacs, and spermaceti may be used both to direct beams of sound and to amplify greatly the energy content or power of the emitted trains of clicks. A key to this likely function is the thick muscle layer of the spermaceti organ. Calculations show it capable of pulling the entire organ backwards with a force of about 10 tons. That much muscle energy compressing the air spaces in the diverticuli could produce a pneumatic sound generator capable of creating high intensity sounds. Scientists theorize that the power of clicks emitted by this system debilitate squid and other organisms, which are then sucked into the mouth of the feeding Sperm whale. Sperm whales may thus feed by using a sonic death beam, which they accurately aim in the dark depths of the ocean.

Sperm whales commonly dive to great depths where the hydrostatic pressure exceeds 100 atmospheres. Their feeding dives often last 30 minutes to an hour. This physiological feat could not be duplicated by human divers wearing scuba or fixed helmet diving gear. This is because humans must breathe air continually, and under pressure, when at depth. Nitrogen dissolved in the blood under pressure will bubble like soda water when the pressure is lessened as the diver returns to the surface. Whales, however, only breathe at the surface. When they descend their lungs collapse at about 100 meters, and the gases in them are forced into the system of nasal sinuses where access to the blood is limited. Little nitrogen thus enters the blood under pressure. When the whale surfaces, its lungs reinflate and (after a long, deep dive) it takes several breaths over a period of a few minutes—"panting"—to repay the oxygen debt it incurred.

Although Sperm whales are fascinating, highly adapted, and magnificent animals, only the mature males are able to succeed in the conditions posed by the Southern Ocean. An odontocete better adapted to this world is the Killer whale. This whale ranges the globe, but it is especially at home in the Antarctic where it can live out its entire life cycle.

Killer Whale

Orcinus orca is an ominous presence in the Southern Ocean. It is the largest member of the dolphin family (Delphinidae), and is one of the most distinctive and brightly marked of all the cetaceans. Adults feature rich black backs and sides, contrastingly set off with white spots on the head and the sides behind the dorsal fin. The pure white of the belly runs from the chin back to the flukes and spreads up over the sides in the caudal region. Unlike most dolphins, Killer whales do not show a pronounced "break," having instead an arched forehead. The large mouth is armed with prominent peg-like white teeth. The females and immature males have large, falcate dorsal fins. The dorsal fin of mature males is exceptionally tall and sail-like, its triangular form projecting as much as 2 meters above the back of a 10-meter whale. The pectoral fins or flippers are also exaggerated in the bulls, and grow to sizes that seem unwieldy.

Killer whales are social animals. They travel, feed, and rest in groups or packs that include both sexes and all age classes. Like Sperm whales they show sexual dimorphism, with males being considerably larger than females. This suggests a polygynous mating system that seems not to have been studied as well as that of the Sperm whale.

These sleek, fast-moving whales are a familiar presence in the waters near the Antarctic Peninsula. They penetrate to leads and pools (polynyas) in the pack ice, and have even been recorded deep in the pack during the dead of the southern winter. It is probably true, however, that in winter most Killer whales move to the north, following those prey that make up their major food supply.

Unlike many cetaceans, Killer whales come close inshore where they seek out sheltered coves for rest and behavior that we interpret as play. An entire group of these whales, resting in shallow water sometimes dive simultaneously, surfacing in the same place after five minutes. Individuals

Male and female Killer whales

nudge one another, or roll over on their backs and slowly sink beneath the surface, making whistling sounds that differ from echolocating clicks or pulsed calls made when the animals are active or hunting.

During the reproductive season courtship features breaching and both vertical and horizontal copulatory positions. Gestation takes about 12 months. Ovulation shortly follows birth, so the females reproduce on an annual basis. At birth the calf is about 2 meters in length. Females reach sexual maturity at about 5 meters in body length, males at about 7 meters.

The real distinction of Killer whales is their diet. They are the only cetacean that routinely attacks and devours warm blooded animals including both birds and mammals. Although in many parts of the world they are principally fish eaters, in the Southern Ocean their diet is heavily flavored with other whales and seals. In one study of their feeding habits, 49 stomachs were examined. Of these 5 were empty. Of the rest Minke whales were found in 84 percent; seals in 45 percent; fish in 7 percent; and squid were found in only 2 percent of the stomachs.

Killer whales hunt cooperatively, ganging up on larger prey such as Minke whales and the juveniles of other baleen species. Their attacks

Killer whales spy hopping

result in the consumption of some blubber, but the lips and tongues of the large whales are especially favored. Antarctic whalers frequently reported that Killer whales would attack and bite chunks out of dead whales that were being towed to factory ships or whaling stations for processing.

Other favorite prey near the Peninsula are seals and penguins. Many scientists working in the Antarctic have reported cases of two or more Killer whales attempting (and sometimes succeeding) in dislodging seals from ice floes where they had taken refuge from the whales. This behavior involves the whales smashing through thin ice from below the seals, or physically tipping the floe back and forth until the seal slides into the water (and into the mouth of a waiting whale). Killer whales are good at locating seals and penguins on ice by raising their heads well out of water as they hang vertically near the surface. This "spy hopping" behavior means that their vision is good out of water. More than one scientist in the Antarctic has been frightened when out on the ice by being eyeballed by curious or hungry Killer whales, but reports of attacks on humans by these animals do not exist in the Antarctic, or elsewhere in the world for that matter.

The diets of all cetaceans in the Southern Ocean, mysticete and odontocete alike, consist only of animal matter—material that is generally easier to digest than the glassy or woody material eaten by herbivores. But the digestive tract of these whales is surprisingly complex—quiet similar, actually, to the conditions found in terrestrial herbivores such as ungulates. Whale stomachs are subdivided into at least three chambers, with the last two being glandular and capable of secreting the gastric juices that initiate digestion (salivary glands are vestigial or absent in whales).

The anterior-most chamber of the stomach is lined with a thick, horny material, and could have a grinding function reminiscent of a gizzard. Some whales, Bottlenose whales for example, are commonly found with stones in their stomachs that they apparently pick up from the bottom. Certainly mysticete whales cannot masticate their food, and odontocetes don't chew either, using their teeth only to tear large prey. Perhaps the complexity of the stomach atones for the failure to process food in the mouth.

In addition to complex stomachs, most whales have exceptionally long intestines—again, similar to the intestinal proportions of herbivores. Records exist, for example, of a 15.2 meter (50 foot) Sperm whale with an intestine 792 meters (2,600 feet) long. Certainly this intestine was relaxed after death and considerably stretched—but still, this is more than three-quarters of a kilometer! The function of this intestinal exaggeration can only be guessed at. Perhaps it is a persistent feature retained from the herbivorous, ungulate terrestrial ancestors that produced the modern whales.

SEALS

If predatory whales near the Peninsula are interesting in the details of their adaptive biology, so are the fish- and squid-eating seals of this region. Four species of those seals live here that are diverse in habitat selection and the ways in which they make their livings.

Southern Elephant Seal

Although the first of these seals—the southern elephant seal, *Mirounga leonina*—breeds over a wide range in the subantarctic islands, it com-

Southern Elephant Seal

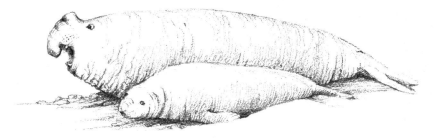

Bull and cow Elephant Seals

monly reaches the Antarctic Continent, and is extending its breeding range through the Peninsula. This animal has a different biology from the rest of the antarctic seals. It is even more sexually dimorphic than the Antarctic fur seal, with bulls being as much as seven times the size of mature cows. Indeed, this is the most extreme degree of sexual dimorphism of any mammal. As in the fur seal this is a reflection of its polygyny, with males competing with each other for the favors of groups of females.

Southern elephant seals select exposed beaches of cobble or boulder for their breeding. Although basically shore-breeding seals, they also haul out on fast ice where they can reproduce successfully. The cows arrive first and gather in groups that are then appropriated by the later arriving bulls. Unlike fur seals, southern elephant seal bulls compete over the groups of females, not a piece of territory to which they then attract potential mates. The larger, more successful bulls generally lay claim to the largest groups of cows. These "beachmasters" may be as much as 4.5 meters in length, and at the peak of condition, weigh over four tons. As in other polygynous species like the Sperm whale, sexual maturity and social ma-

"The male elephant seals would rear up and make bellowing sounds and snap at each other, grabbing skin around the neck occasionally. Most of them had scars and some open sores. They are huge and unwieldy-looking on land."

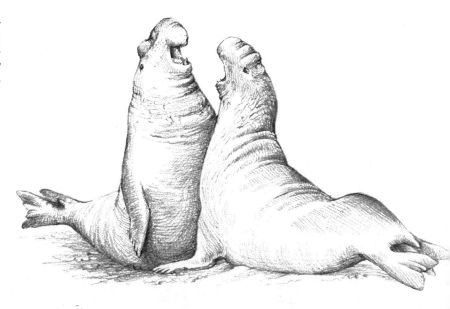

turity are delayed so the bull southern elephant seals are not reproductively successful until they attain an age of at least nine to fourteen years.

Both the cows and the bulls undergo long periods of fasting during their reproductive period and in the molt which follows. To that purpose they lay down exceptionally large stores of blubber to see them through this lean and metabolically active time. It is for that reason that they were extensively harvested in the nineteenth century as adjuncts to both the whaling and sealing fisheries. The degree of exploitation of them was severe, although not as excessive as for the Antarctic fur seal; and although depleted, their stocks were not as close to extinction. Like the Antarctic fur seal, they have recovered. Because they are primarily fish and squid feeders, however, the demise of whale stocks has not had the stimulatory effect on their present populations in the same way that it affected both fur seals and crabeater seals. The population of southern elephant seals (that now number about 800,000 animals in the Southern Ocean) seems now to be stable, or even decreasing slightly in some areas.

After the rigors of reproduction that include aggressive competition between males; and parturition, lactation, and mating in the females, both sexes move to the water for a month of recovery before returning to land to molt late in the summer. Unlike the molt of other antarctic seals, this is a "catastrophic" one, with hair and epidermis shed in large patches and sheets. During this period, which may last a month, the elephant seals retreat inland, and rest in meltwater pools or ponds that they quickly foul with both their shed hair and skin, and with their excrement. The sight is not a pretty one, and the odor rivals that of large penguin colonies on warm days. In some places elephant seals have polluted small ponds to the point that they show the classic symptoms of eutrophication, and have acquired unique ecological characteristics.

Southern elephant seals are deep divers. They eat fish, squid, and oc-

Southern Elephant Seal calf

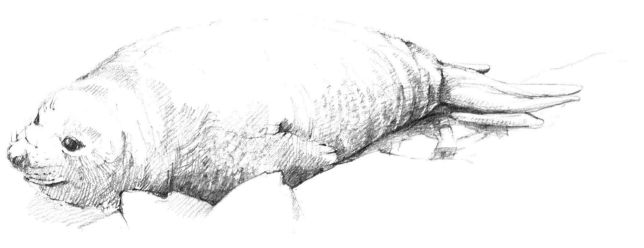

''The elephant seals on the beach are molting and slobbering and making belching noises at each other. They are lying everywhere here—huge mammoth, bellowing bags of blubber. The stench is horrible—both from the orange-brown muck they're lying in and their steaming breath when they open their slimy, foamy mouths to belch. I can't think of a more disgusting animal. Some people have nicknamed them 'pig seals,' considering it a more appropriate term—but what an insult to pigs!''

topuses in the main, and they are known to dive as deep as 300 meters (and likely much deeper) to catch this food. The adaptations for these feats are similar to those known for the Weddell seal, an easier subject to study. Southern elephant seals' eyes have retinas with visual pigments that seem especially well suited to catching prey in the dark depths at which they hunt. This pigment absorbs light wavelengths best in the spectrum produced by the bioluminescence of the squids that they eat.

The life of southern elephant seals away from the breeding beaches and molting retreats, like that of many of the antarctic seals, is little known. Presumably they move away from the highest latitudes during the winter, and range to the ice edge during the summer. Their high seas behavior, and details of possible migrations remain enigmatic.

Ross Seal

In a similar fashion, not much is known about the biology of another cephalopod specialist in the true Antarctic, the Ross seal *(Ommatophoca rossii)*. Unlike the southern elephant seal this is a fully pelagic animal that hauls out only on ice floes in the denser elements of the pack. Living only in the most inaccessible regions (away from the ice edge as well as the fast ice near shore), the Ross seal was long though to be a rare and elusive creature. Since 1970, however, aerial surveys found this pinniped to be more abundant than previously thought. It is a solitary seal, not living in dense aggregations, but rather singly or in pairs hauled out on the dense larger floes of the inner pack.

Ross Seal

Population estimates for the Ross seal hover around 220,000—about that of the lower limit for leopard seals. Much remains to be learned about the general and reproductive biology of this animal. It is a handsome seal, displaying a definite two-toned appearance (dark above, light below) when hauled out on the ice. When approached on the ice this seal raises its head and impresses the observer with its large eyes. This last feature is probably an adaptation for feeding in the dimly-lit deep depths at which it apparently feeds, for its diet consists mainly of squid and other cephalopods. This seal features small, sharp teeth—useful in eating squid. They eat some of the larger squid species, with mantle lengths nearly three quarters of a meter.

Few Ross seal pups or fetuses have been studied. The breeding season is probably later than that of crabeater or Weddell seals, perhaps similar to the reproductive timing of leopard seals.

Because they prefer the denser pack ice areas, Ross seals may not be attacked by Killer whales and leopard seals to the same extent as crabeater seals. Their small populations, rather, reflect the scarcity of foodstuffs at their trophic level; or our failure to measure adequately their numbers.

Weddell Seal

If our understanding of the biology of Ross seals is comparatively slight, we know a great deal more about the life and physiology of its inshore neighbor, the Weddell seal *(Leptonychotes weddelli)*. This animal, which breeds in the fast ice region of the Peninsula and Antarctic Continent, is especially accessible because it hauls out on beaches, and through tide cracks and breathing holes in fast ice that can be visited on foot or by snow vehicle. In addition, it is not usually aggressive or fearful of man. With a few noteworthy exceptions it tolerates close approach and handling by scientists. This benign behavior may stem from the fact that its spring habitat is well inshore of the more open waters patrolled by Killer whales and leopard seals. It is the most southerly year-round resident of the Antarctic, excepting man.

"Weddell seals let you come right up to them. Funny animals—they can hardly move on land. It's like a huge lump of dough that has to undulate with great effort to flop its way along inch by inch. It looks like a lot of work. We accidentally startled one by coming over a rock. It opened its mouth wide and backed up slightly with its undulating motion, every bump rippling across its flesh like jello."

Male Weddell Seal under water

Because of their docility and accessibility Weddell seals can be approached and fitted with all sorts of instrumentation and tags. Thus we know a good deal about their diving capabilities, population dynamics, diet, and physiology.

Weddell seals are fish eaters, fish counting for more than 50 percent of its stomach contents. Various invertebrates (35 percent) and squid (11 percent) are also frequently eaten. This food is caught at depths ranging down to 600 meters. On average more than 20 kilograms of food is taken at a single meal. Because the seals fish from fairly permanent holes near the tide cracks that delimit the breeding areas through the spring and early summer, local populations of fishes like Antarctic cod can be depleted in short order.

In their reproductive behavior, Weddell seals must rank among the most unusual of mammals, or any other vertebrate for that matter. The seals are apparently polygynous with males mating with more than one female. Unlike most polygynous mammals, however, males are slightly smaller than the cows, weighing 350 to 450 kilograms in the spring. The qualifier, "apparently," is because mating occurs beneath the ice and water surface where the males establish and defend underwater territories along the tide cracks and near the breathing holes that give them access to air. While females haul out to rest, give birth, and nurse their pups, the males almost never emerge during the breeding season, patrolling instead the three dimensional territories that the cows must traverse to reach the ice surface.

Early in the breeding season territories average abut 6,000 square meters of surface area, diminishing later on to half that size in a manner analogous to that seen in Antarctic fur seals. The territorial males are highly interactive, becoming emaciated as the season progresses and suffering numerous cuts and wounds. Unlike elephant and fur seals that become scarred about the head and shoulders from their territorial conflicts, Weddell males also get bitten around the genital region and tail, suggesting that a lot of chasing goes on in addition to head-to-head conflict.

The smaller size of the male is advantageous in that small males are more maneuverable and agile, giving them better ability to mate in the challenging watery medium, as well as to defend effectively their positions along the tide cracks.

Studies of the population dynamics of these seals show that as many as half of the adult females forego reproduction in any given year. Although this suggests that breeding areas are limited by the extent and number of tide cracks, experimentally displaced pregnant females successfully rear young in more dense aggregations than are usually seen in the wild. Moreover, the breeding females normally associate with non-breeders, although the latter are more transient than the former. Sexual maturity in Weddell seals is reached in from three to six years in both females and males. Social maturity in the males, however, does not occur until the age of seven or eight. Adolescent seals are noticeably absent from the breeding aggregations, and spend much of their time seaward and to the north of the fast ice regions. Indeed, immature and non-breeding adults are common in the vicinity of Palmer Station on Anvers Island.

Unlike most other seals of the Peninsula, Weddell seals do not seem to be especially migratory, probably spending the long winter in the same fast ice regions in which they breed. Here they depend upon tide cracks, trapped air pockets under hummocked ice, and breathing holes that they keep open for access to air. To this purpose the seals are equipped with protruding canine and incisor teeth that they use to chisel open breathing

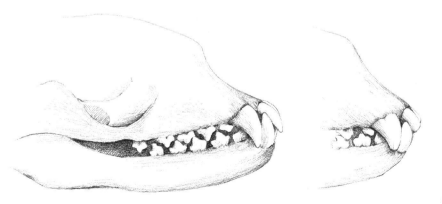

Procumbant teeth of Weddell Seal, comparing young (left) with old, worn (right) dentitions

holes in the winter ice. Although they have no predators in the fast ice environment, progressive tooth wear is one factor curtailing their lifetimes. When the teeth are no longer competent as ice chisels, death by drowning may occur. Few Weddell seals live to the age of twenty—an age often attained by other seals that spend their winters in more open waters.

Weddell seals are the premier underwater artists of the pinniped world. Their adaptations to the underwater life are extreme, allowing them to remain submerged for periods in excess of one hour. Microcomputer and other monitoring experiments show two distinct diving patterns in these seals. The dives that are longest in duration (more than 20 minutes), are usually the shallowest, with the seals remaining in visual contact with the overlying ice. These are exploratory dives during which the animals make excursions as far as 12 kilometers away from their point of departure, searching, we can suppose, for additional breathing holes. The second diving pattern is for feeding and features dives of short duration (usually less than 15 minutes), but of great depth. These feeding dives commonly extend from 200 to 400 meters, with the deepest ones going to 600 meters or beyond. In some seasons (spring, for example) most of the diving activity occurs during the daytime.

Like the deep-diving Sperm whale, Weddell seals avoid nitrogen narcosis and the ''bends'' by having flexible rib cages that allow the lungs to collapse at moderate depths before much nitrogen can enter the blood. Experiments in which blood was intermittently sampled from diving seals demonstrate that the collapse of the lungs occurs at depths of from 40 to 80 meters. In addition to this adaptation, these seals also show dramatic changes in the blood circulation patterns when forced to dive in the laboratory. Oxygen is conserved by shunting most of the active blood flow away from the viscera and muscle to the brain. When the animals resurface, blood is delivered to the muscle, and the oxygen debt incurred during the dive can be repaid. The same mechanism probably occurs in the wild as well.

All in all the Weddell seal is a most incredible animal, showing a set of physiological, behavioral, and anatomical adaptations for living in one of the most extreme environments imaginable. It has survived because of its remoteness and lack of interest from the only predator that can reach it—man.

Leopard Seal

The final seal to consider here is that dietary jack of all trades, the leopard seal *(Hydrurga leptonyx)* whose krill feeding adaptations were considered in chapter 7. The leopard seal, however, is also a predator par excellence, and this aspect of its behavior will be considered now.

Although the recorded diet of leopard seals approximates 50 percent krill, this food source is utilized more extensively in winter, and less often in the spring and summer. It is during these latter seasons that penguins, especially Adélies, congregate for breeding, and the abundant crabeater seals are pupping. Some leopard seals, especially mature males, specialize on these food sources at this time of year.

In the well-studied Adélie penguin rookery at Cape Crozier in the Ross Sea, for example, the 200,000 penguins are annually harassed by three to six offshore leopard seals. These seals find a ''land of plenty'' during the three to four months that they are in attendance. Generally the seals employ several strategies to capture the Adélies. Early in the season when the penguins must walk long distances over fairly thin ice to reach the rookeries, the seals will attack the walking penguins by smashing through the ice from below. Although fewer than 10 percent of these attempts are successful, the seals still manage to wreak psychological havoc among the penguins, the response of which to these attacks is usually to remain stock still until long after the seal has departed.

Later in the season, the leopard seals patrol the beaches and ice fronts to which the foraging adult penguins must return to take their turns at incubation chores. When the seals spot returning penguins they submerge and attack the birds as they reach the beach. In cases when the tide is low and the penguins must repeatedly attempt to leap as much as three meters up the exposed ice front, the seals have easy pickings.

The most plentiful time for the seals comes when the penguin chicks first take to the water. They are clumsy swimmers at this time, completely

''Leopard seals have reptilian looking forms with massive heads and jaws and an evil eye.''

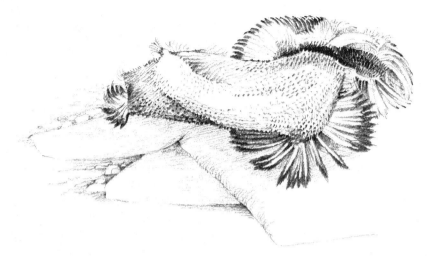

"I also found the remains of a penguin—the skin turned completely inside out. I imagine it was the victim of a leopard seal."

incapable of diving. The seals are then able to feed at will, with more than 90 percent of the attacks being successful. All told, in the breeding season at Cape Crozier, as many as 4,800 adult penguins (2.4 percent of the population) are killed by leopard seals. Mortality of the chicks amounts to about 1,200 birds, or 1.2 percent of the chick population.

When a leopard seal captures a penguin it quickly snaps the bird through the air with a powerful shake of its head. This is often enough to split the penguin's skin and fling it aside. But even if the skin does not come away, the seal will eat the penguin whole—feathers and all. Sometimes a leopard seal repeatedly dashes the unfortunate bird's carcass against the surface of the water, progressively dismembering it and tearing away pieces of flesh that are then swallowed. Although the leopard seal's teeth are admirably suited for straining krill from the water, they are less efficient at severing flesh and bone.

The other significant prey of the leopard seal are the young of the crabeater seal. As we saw in chapter 7, this is the most abundant seal in the world, and although the breeding adults are widely dispersed throughout the looser elements of the pack ice, leopard seals do well at capturing the pups.

Crabeater seal pups can be pursued by leopard seals on the ice floes before they are weaned. Most often, however, the young seals are attacked in the water. Seals younger than a few months almost never are found with the characteristic parallel scars from leopard seal attacks, suggesting that the youngest crabeaters do not survive to show the scars. As they grow they become more formidable prey and are more likely to escape the advances of the predator. These seals carry scars. By the time the young crabeaters are approaching a year of age, the percentage found with fresh wounds diminishes to near zero, meaning that they are now too large to be attacked by the leopard seals. The degree of predation must be high, for as many as 80 percent of all crabeaters carry with them the scars that indicate a hostile encounter with a leopard seal.

The details of the attack pattern of leopards on crabeaters are not known, but carcasses of crabeater pups have been found with the blubber and flesh peeled away from the hind ends—presumably by leopard seals.

Despite the prevalence of leopard seals as noteworthy predators in the region of the Antarctic Peninsula, their numbers are nowhere dense. Probably no more than 220,000 leopard seals exist at the present time in the Southern Ocean, and the extreme degree of their dispersion is a sign of the relative dearth of food for a top carnivore in this, or really any environment.

Beyond the slim pickings for the tertiary level of consumers in the antarctic ecosystem, food gets very scarce indeed. What is left is taken care of by the parasites that infest many of the birds, mammals, and fishes of the Antarctic. The remaining energy goes to the detritus food chains that are populated by micro-organisms such as fungi and bacteria. The importance of bacteria in the Southern Ocean has yet to be firmly established, but it is likely to be high.

ADDITIONAL READING

Brown, S. G. and C. H. Lockyer. 1984. Whales. In R. M. Laws, ed., *Antarctic Ecology*, 2:717–781. London and Orlando, Fl.: Academic Press.

Dearborn, J. 1965. Food of Weddell seals at McMurdo Sound, Antarctica. *Journal of Mammalogy*, 46(1):37–43.

Falke, K. J., R. D. Hill, J. Qvist, R. C. Schneider, M. Guppy, G. C. Liggins, R. W. Hochachka, R. E. Elliott, and W. M. Zapol. 1985. Seal lungs collapse during free diving: Evidence from arterial nitrogen tensions. *Science*, 229:556–558.

Gaskin, D. E. 1982. *The Ecology of Whales and Dolphins*. London and Exeter, N.H.: Heinemann.

Goodall, R. N. P. and A. R. Galeazzi. 1985. A review of the food habits of the small cetaceans of the Antarctic and Subantarctic. In W. R. Siegfried, P. R. Condy, and R. M. Laws, eds., *Antarctic Nutrient Cycles and Food Webs*, pp. 566-572. Berlin: Springer Verlag.

Jehl, J. R., W. E. Evans, R. T. Awbrey, and W. S. Drieschmann. 1980. Distribution and geographic variation in the Killer whale *(Orcinus orca)* populations of the antarctic and adjacent waters. *Antarctic Journal of the United States*, pp. 161–163.

King, J. E. 1983. *Seals of the World*. 2d ed. Ithaca, N.Y.: Cornell University Press.

Kooyman, G. L. 1981. *Weddell Seal: Consummate Diver*. Cambridge and New York: Cambridge University Press.

Matthews, L. H. 1978. *The Natural History of the Whale*. New York: Columbia University Press.

Müller-Schwarze, D. 1984. *The Behavior of Penguins Adapted to Ice and Tropics*. Albany: State University of New York Press.

Norris, K. S. and B. Mohl. 1983. Can odontocetes debilitate prey with sound? *American Naturalist,* 122(1):85–104

Ridgway, S. H. and R. J. Harrison, eds. 1981. *Handbook of Marine Mammals.* Vol. 1: *The Walrus, Sea Lions, Fur Seal and Sea Otter;* Vol. 2: *Seals.* New York: Academic Press.

Siniff, D. B. and J. L. Bengston. 1977. Observations and hypotheses concerning the interactions among crabeater seals, leopard seals, and Killer whales. *Journal of Mammalogy,* 58(3):414–416.

Siniff, D., D. P. DeMaster, R. J. Hofman, and L. L. Eberhardt. 1977. An analysis of the dynamics of a Weddell seal population. *Ecological Monographs,* 47:319–335.

Siniff, D., S. Stone, D. Reichle, and T. Smith. 1980. Aspects of leopard seals *(Hydrurga leptonyx)* in the Antarctic Peninsula pack ice. *Antarctic Journal of the United States,* p. 160.

Slijper, E. J. 1979. *Whales.* 2d ed. Ithaca, N.Y.: Cornell University Press.

Testa, J. W., D. B. Siniff, M. J. Ross and J. D. Winter. 1985. Weddell seal—Antarctic cod interactions in McMurdo Sound, Antarctica. In W. R. Siegfried, P. R. Condy and R. M. Laws, eds., *Antarctic Nutrient Cycles and Food Webs,* pp. 561–565. Berlin: Springer Verlag.

Thomas, J. A. and D. P. DeMaster. 1983. Diel haul-out patterns of Weddell seal *(Leptonychotes weddelli)* females and their pups. *Canadian Journal of Zoology,* 61(9):2,084–2,086.

Thomas, J. A. and V. B. Kuechle. 1982. Quantitative analysis of Weddell seal *(Leptonychotes weddelli)* underwater vocalizations at McMurdo Sound, Antarctica. *Journal of the Acoustical Society of America,* 72(6):1,730–1,738.

Thomas, J. A. and I. Stirling. 1983. Geographic variation in the underwater vocalizations of Weddell seals *(Leptonychotes weddelli)* from Palmer Peninsula and McMurdo Sound, Antarctica. *Canadian Journal of Zoologoy,* 61(10):2,203–2,212.

CHAPTER 10

THE FUTURE

Only 140 million years ago (about 0.03 percent of the earth's history) the land mass that we now know as Antarctica was a vastly different place. Part of the disintegrating Gondwanaland, it was about to begin its most recent slide to the South Pole and there become surrounded by its circumpolar ocean. That latter event, occurring 20 to 30 million years ago, set the stage for the formation of its ice cap, which, as we have seen, obliterated a rich terrestrial fauna and flora, and created unique meteorological and oceanographic conditions that persist today.

Within only the last million years of this history the ice sheet has expanded and receded several times. With each expansion the oceans of the world subside, and great amounts of rock and gravel are bulldozed by the ice out onto the antarctic continental shelf. As the ice recedes—as it has done for the past 18,000 years—the oceans recover their volume, and ice-free land appears, ready to be colonized by plants and animals emerging from refugia, or drifting in from the north.

In the measured pulse of geological time the earth and the continents that populate its surface are not constant. Antarctica has changed in the past, and as the earth wobbles in its course around the sun, it will change in the future. In only the past 100 years, however, a new element has been added to the factors that can forge antarctic change—Mankind.

In the aboriginal state humans are not a physiological match for the

climate and insularity of modern Antarctica—until recently the only un-inhabited continent. Our technical prowess, driven by curiosity and competitiveness, now threatens to overwhelm the natural grandeur of this land, and forge changes of different dimensions more rapidly than would otherwise occur. What could these changes be? And what forces exist that might promote or retard them?

FAUNA AND FLORA

The snow and ice-free terrestrial environment of Antarctica is scarce, with most of the available habitat occurring on the Peninsula. As we saw in chapters 3 and 4, the sparse community of plants and animals that reside here consists mostly of mosses, lichens, microbes, and a few small invertebrates. These organisms grow slowly—indeed, many of the moss plants and lichen colonies here today were probably alive when Antarctica was first discovered more than 150 years ago. In ecological terms this terrestrial community is unproductive, with low species diversity—a measure both of the number of species and their relative frequencies of occurrence. There are, so to speak, very few eggs here in just a few baskets. This kind of habitat is at risk, both from incremental changes of natural origin, and those stemming from the callous hand of man.

In the first instance the important change will be one of climate—colder times will obliterate the present habitat with ice; warmer years will allow

the invasion of a more diverse flora and fauna, probably similar to that found today on subantarctic islands like South Georgia, Kerguelen, and Macquarie. The effect of man's activities, however, can be more direct and disastrous. Construction of field stations, introduction of exotic species, biological sampling, waste disposal, toxic pollutants, tourism, and foot and vehicular traffic can all have detrimental effects on the native flora and fauna. With few exceptions, the field stations established on the Peninsula (there are presently nineteen active ones, supported by seven different nations—historically, between thirty-five and forty sites have been occupied) are built on the level ice-free surfaces so much at a premium here for terrestrial life.

Humans in Antarctica have a rich history of irresponsible trash disposal. Solid wastes, in particular, are enduring features of the Peninsular landscape, and many stations had their surroundings littered with cans, barrels, and tins of all descriptions. That waste remains in few cases (it is even protected at historical sites). Occupied stations usually abide by agreed conventions that call for (a) the subtidal dispersal of liquid and sanitary wastes, (b) the disposal at sea of non-combustible solid wastes, and (c)

Field station

the removal from the Antarctic of rubber, plastic, radioactive, and toxic wastes.

Superimposed on the fragility of the resident terrestrial community of the Antarctic Peninsula is the seasonal complement of sea birds and seals that use the land for nesting and hauling out purposes. Human proximity to traditional rookeries and the pollution of nearby areas has, in some instances, adversely affected such nesting and resting areas. In the past, too, explorers and scientists routinely relied on penguins, seals, and bird eggs for food—both for themselves and their sledge dogs.

The populations of sea birds and seals that use the Antarctic Peninsula are probably more affected by activities that go on in the sea than on land. This, after all, is the basis of their lives. There is a history of disruption of the marine environment by man, and the ominous threat of more in the future.

Marine Exploitation

The first great wave of antarctic exploitation took place in the 1820s with the discovery and rapid demise of the populations of Antarctic fur seals

Penguins nesting on coal bags

Elephant seal

that were found in the South Shetland Islands and elsewhere on the Peninsula's archipelago. In the span of only two years hundreds of thousands of these animals were slaughtered, bringing them close to extinction. Their numbers recovered slightly in the next fifty years, and in 1870 a briefly revived fishery once again eliminated them as a commercial resource.

Following the depletion of fur seal stocks, the southern elephant seal became the next target of exploitation. Throughout the nineteenth century and well into the twentieth it was harvested for the oil rendered from its blubber, often as an adjunct to the whaling business. Although southern elephant seals were hunted ruthlessly, several factors kept them from being brought as close to extinction as the fur seals. The sexual dimorphism in this species made the bulls much more valuable than the cows. These great animals were so large that it was not easy to kill them, strip the blubber, and render it. Also, the intensity of the fishery was never what had developed for the more valuable fur seal. It was possible then for authorities to institute regulations to extend the harvest of these animals (which continued on South Georgia as recently as 1964).

The other great historical exploitation of the Southern Ocean involved, of course, whales—particularly the huge rorquals that feed on krill here during the summer. Although whalers had long penetrated the fringes of the Antarctic to search for Sperm and Right whales, the invention of exploding harpoons and steam-driven whale catchers made the hunt for Blue, Fin, and Sei whales possible and profitable.

By 1904–1905 commercial whaling came to the Antarctic on a grand scale with the establishment of a permanent whaling station on South Georgia by the Norwegian whaler and explorer C. A. Larsen. This station was a prototype for others to be founded on South Georgia and elsewhere, even to the South Shetlands. The imposition of a duty on South Georgia whale oil by the British (this tax supported the *Discovery* expeditions that so increased our knowledge of the Southern Ocean) forced the

Whale bones on beach

Norwegian whalers into factory ships where the rendering of whale oil was carried out on the high seas. Larsen ultimately passed away in 1925 on a factory ship in the Ross Sea while attempting to open a new whale fishery there.

The depletion of the great whales in the Southern Ocean is a sad tale of international conflict, greed, and mismanagement of a resource. The slaughter of these whales abated somewhat during the years of World War II, but inexorably their stocks were depleted, first the Blue, then the Fin, and now the Sei whale. Only the Minke whale maintains stocks of a reasonable size, and even they are now the subject of a whale fishery conducted by Japan and the Soviet Union.

For the Southern Ocean, the great effect of the loss of its whales was the surplus of krill production that these whales once ate. That production has been estimated to be over 150 million tons per year. As we saw in earlier chapters, some of that production seems to funnel into ever larger stocks of other krill-eating animals, such as Adélie penguins, Antarctic fur seals, and crabeater seals. It also presents a resource that is being tested by the fisheries of nations like Japan and the Soviet Union. To be sure, the stocks of krill in the Southern Ocean form a resource far greater than the total fisheries catch in all the rest of the world. Expoitation of it seems inevitable.

MINERAL RESOURCES

Every well-explored continental land mass in the world has proved to have exploitable mineral resources of diverse kinds. There is no reason to suspect that valuable minerals do not exist in Antarctica. Already deposits of coal and iron ore have been identified. But the geological history of this region strongly suggests that other sources of wealth lie here as well. The matching pieces of Gondwanaland that today exist as South Africa and Australia hold ores rich in platinum, chromium, lead, cobalt, nickel, and other rare earth metals. The same geological processes that formed the South American Andes with its rich mineral lodes also operated in the Antarctic Peninsula, and make likely the discovery of copper, zinc, silver, gold, molybdenum, and other valuable metals here.

The antarctic continental shelf, although narrow and deep, holds the possibility of containing deposits of gas and oil. Indeed, traces of hydrocarbons were found in experimental drilling in the Ross Sea several years ago. It is only a matter of time, many observers think, before the commercial exploration for and exploitation of possible gas and oil reserves takes place. As we saw in chapter 2, the water present in the great icebergs of the Southern Ocean is a resource coveted by nations of the world that are starved for adequate quantities of fresh water. Already a lot of money has been spent on finding ways to transport icebergs, or their water, to dry areas of the globe.

The hostile, extreme environmental conditions of the Antarctic have hindered both the exploration for these mineral resources and the commercial development of those already found. We can probably expect that this condition will continue for some time to come. The history of human endeavor, however, tells us that if existing exploitable resources become scarce enough, the search for more will be intensified, and their extraction will occur, even in the frozen, ice-covered land of Antarctica.

Quartz crystals

Geologists' field camp

TOURISM

Since the mid 1960s increasing numbers of tourists have visited Antarctica. Many travel to the Peninsula and islands on regularly scheduled cruise ships such as the *Lindblad Explorer* and the *World Discoverer*. Others have flown over parts of the continent on planes departing from Chile, Argentina, and New Zealand. Antarctica is also becoming a mecca for adventuresome private parties that sail or fly there for the purposes of sightseeing, mountain climbing, or scientific investigation. All of these activities result in some degradation of the environment, and are carried out at risk to the individuals involved. But on the other hand, tourism promotes a

World Discoverer

heightened awareness in more people of both the beauties of Antarctica and some of the problems it faces.

The repeated visits by tourists to the nesting areas of penguins and other sea birds can be disruptive, especially at the time when eggs or young chicks are being incubated. Despite regulations to the contrary, cruise ships are known to pump bilges in the sensitive near-shore regions of the continent, fouling the water with oil and other waste. Air traffic involving tourists has resulted in unfortunate accidents, the most serious of which took 257 lives when an Air New Zealand DC 10 crashed on Mount Erebus near the Ross Sea. In the summer of 1985/86 a Chilean airplane went down on King George Island, killing all people on board.

Because accidents involving tourists are likely, it has become the position of several countries (including the United States) doing business in the Antarctic not to encourage this kind of activity. Private parties in distress turn to scientific programs for aid and rescue, costing the programs much lost time and resources.

TERRITORIAL CLAIMS

From the time of its initial discovery and exploration various territorial claims have been made on Antarctica and its related islands. Among the earliest of these was the United Kingdom's in 1908 that set up the Falk-

Map of territorial claims

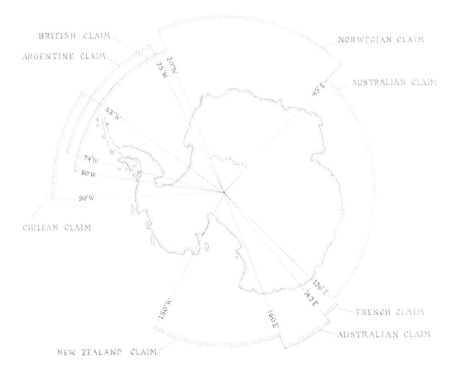

land Island Dependencies. Other nations making territorial claims here include Argentina, Chile, New Zealand, Australia, France, and Norway. Several more have conducted explorations in the Antarctic, but for one reason or another have not pressed territorial claims. These include Japan, West Germany, Belgium, the United States, and Russia. Both the U.S. and the U.S.S.R. have been adamant in neither claiming territory nor recognizing the territorial claims of other countries.

The most contentious claimants to parts of Antarctica are the United Kingdom, Argentina, and Chile, for their various claimed regions overlap and all include the Antarctic Peninsula. The fractiousness resulted in shots being fired in 1952, and more serious encounters in 1982 during the Falkland Islands dispute.

The territorial claims, according to international law, can possibly be legalized by maintaining an ''effective occupation'' of the territory. For some nations this is a stimulus to establish and man scientific stations in the Antarctic. Argentina and Chile, in particular, have engaged in a program of establishing colonies of people, including families that live year-round in certain stations on the Peninsula and in the South Shetland Islands. The first baby born in the Antarctic was baptized in 1978 at the Argentinian Esperanza Base near the northern tip of the Peninsula. Since 1984, the families of workers at the Chilean Teniente Marsh Base on King George Island have lived year-round there.

The Antarctic was generally regarded prior to World War II as an interesting, but hostile and difficult place to be. Territorial claims really were not pressed. Following the war, however, technological advances made this continent more accessible, its declining whale stocks more sought after, its strategic importance more real, and its scientific value better appreciated. Accordingly, international discussions were set in motion that resulted in an extraordinary agreement—the Antarctic Treaty.

ANTARCTIC TREATY

In the 1950s the scientific community of meteorologists and geophysicists was interested in using Antarctica as a platform from which to make observations of polar climate, space, and the earth's magnetic field. An international geophysical year (IGY) was proposed for 1957–1958 to accomplish this multidisciplinary, multination cooperative task. The scientific results obtained, and the cold-war-breaching cooperation of using Antarctica in a unified way for science provided the impetus for a more general political agreement. Accordingly, in 1959, twelve interested nations (Argentina, Australia, Belgium, Chile, France, Japan, New Zealand, Norway, South Africa, the United Kingdom, the United States, and Russia) signed a treaty, the Antarctic Treaty, that formally went into effect on

June 23, 1961. The basic agreement, which contains fourteen articles, establishes a number of constraints designed to foster international cooperation, and, most importantly, protect Antarctica and its resources from damage and over-exploitation.

Polar Star

The essential points of the treaty are as follows: That Antarctica will be used for peaceful, nonmilitary purposes only; that there will be freedom of scientific investigation including the exchange of personnel and observations; that while the treaty is in force territorial claims will be held in abeyance; that nuclear explosions and disposal of radioactive wastes are prohibited; that consultative meetings among the treaty nations will be held periodically, especially to further the preservation and conservation of living resources; and that activity by any country that is contrary to the treaty will be discouraged by the member nations.

Since the Treaty has been in effect consultative meetings have been held at two-year intervals. The original dozen member states have been joined by four more (Poland, Brazil, West Germany, and India), and an additional 13 have acceded to the Treaty (Bulgaria, Czechoslovakia, Denmark, East Germany, Hungary, Italy, Netherlands, Papua New Guinea, Peoples Republic of China, Peru, Spain, Rumania, and Uruguay). The consultative meetings have made substantial progress in the protection of living resources. For example, the Convention for the Conservation of Antarctic Seals, ratified in 1978, called for the complete protection of Ross, fur, and elephant seals, and set catch quotas on the remaining species. More recently, in 1982, The Convention on the Conservation of Antarctic Marine Living Resources came into effect. This convention allows the Scientific Committee on Antarctic Research (SCAR)—made of up scientists from treaty nations—to conduct the research that will allow the establishment of guidelines for harvesting renewable living resources such as krill.

The Antarctic Treaty was originally designed to be in force for an "experimental" thirty years. When that period is up, in 1991, the member nations can decide, by majority, to make amendments to the treaty. Within the following two years, however, any member can withdraw from the treaty two years after giving notice that it will not abide by the provisions therein. The Antarctic Treaty, therefore, could become unworkable between 1993 and 1995.

The major stresses facing the Antarctic Treaty don't seem to be directly associated with the competition for exploiting the living resources of the Antarctic, rather the likelihood of finding substantial mineral or hydrocarbon resources looms much larger. With the prospect of finding oil and gas under the sea floor, and valuable metals and other minerals in the continental crust, there has come a renewed interest in asserting the territorial claims that the treaty holds in abeyance.

Most recently, some of the poorer, non-industrialized countries of the world have questioned the right of the Treaty nations to hold for themselves the potential wealth of Antarctica, arguing that, like the resources of the deep sea floor, it should be "the common heritage of all mankind." They have brought this question before the United Nations, which is now holding debate on Antarctica and the ownership of its mineral and other resources.

Yet another point of view is being expressed by environmental groups such as the Antarctica Project, the Southern Ocean Coalition, and Greenpeace. They advocate such points as moratoriums on krill fishing and mineral exploration, and the establishment of an "Antarctic Environmental Protection Agency."

It is clear from all of this activity and concern that the future of the Antarctic is no longer firmly in the hands of the few nations that historically have explored and supported research there. The debate has widened, and more eyes are watching. This could be good, for the treasures that Antarctica presents to all mankind are unique, and go far beyond the monetary assessment that commerce puts on natural resources, or even the natural laws that science attempts to discern. In its rugged beauty it is the home of animals that still have no fear of man. More than any other place on earth it remains virtually pristine. In a world increasingly overwrought with the mounting anxieties of an exploding human population, Antarctica represents a psychological refuge that all of us can re-

lish. We all, therefore, have an interest in seeing that this marvelous continent is not despoiled by the gluttony of a few.

ADDITIONAL READING

Antarctica. 1985. Sydney: Readers Digest.

Auburn, F. M. 1982. *Antarctic Law and Politics.* Bloomington: Indiana University Press.

Bertrand, K. J. 1971. *Americans in Antarctica 1775–1948.* New York: American Geographical Society.

Block, W., A. J. Burn, and K. J. Richard. 1984. An insect introduction to the maritime Antarctic. *Biological Journal of the Linnean Society,* 23:33–39.

Bonner, W. N. 1984. Conservation and the Antarctic. In R. M. Laws, ed., *Antarctic Ecology,* 2:821–850. London and Orlando: Fl.: Academic Press.

Deacon, G. E. R. 1977. The Southern Ocean: History of exploration. In G. A. Llano, ed., *Adaptations Within Antarctic Ecosystems,* pp. xv-xxxv. Washington, D.C.: Smithsonian Institution.

Everson, I. 1984. Marine interactions. In R. M. Laws, ed., *Antarctic Ecology,* 2:783–819. London and Orlando, Fl.: Academic Press.

Headland, R. 1984. *The Island of South Georgia.* Cambridge: Cambridge University Press.

Kameneva, G. I. and G. E. Grikurov. 1983. A metallogenic reconnaissance of antarctic major structural provinces. In R. I. Oliver, P. R. James, and J. B. Jago, eds., *Antarctic Earth Science,* pp. 420–422. Cambridge: Cambridge University Press.

Parfit, M. 1985. *South Light.* New York: Macmillan.

Parker, B. C., ed. 1978. *Environmental Impact in Antarctica.* Blacksburg: Virginia Polytechnic Institute and State University.

Polar Regions Atlas. 1978. Washington, D.C.: Central Intelligence Agency.

Shapley, D. 1986. *The Seventh Continent.* Washington, D.C.: Resources for the Future.

Splettstoesser, J. F. 1983. Mineral resources potential in Antarctica—Review and predictions. In R. I. Oliver, P. R. James, and J. B. Jago, eds., *Antarctic Earth Science,* p. 413. Cambridge: Cambridge University Press.

Zumberge, J. H. 1979. Mineral resources and geopolitics in Antarctica. *American Scientist,* 67:68–77.

GLOSSARY

Ablation The removal and loss of ice from the ice sheet through melting, sublimation, and the calving of icebergs at the edges.

Acari The group of arachnids to which the mites belong.

Adiabatic A process by which a fluid or gas can undergo pressure and volume changes without change of heat.

Albedo A measure of the reflectivity or "whiteness" of the earth's surface. Higher albedos result in more reflection of sunlight back into space.

Ammonites Fossil mollusks (related to living cephalopods) with characteristic curved and chambered shells. These animals flourished until the end of the Cretaceous—about 65 million years ago. Although most species were small, giant ammonites existed that had shells more than 2 meters in diameter. Their remains are abundant in certain marine fossil assemblages.

Amphipods Laterally compressed, shrimp-like crustaceans, without fused thoracic segments. These are usually benthic dwellers that feed on detritus.

Antarctic Convergence The circumpolar region where the cold, north-flowing antarctic surface water meets warmer, subantarctic surface water. This region (which usually is found near 50° South latitude) marks the northern limit of the Southern Ocean.

Antarctic Divergence Surface water in the West Wind Drift is deflected to the north by the prevailing winds, while that in the East Wind Drift is moved to the south. Surface water is thus moved away from the shear zone between the two wind systems. At this divergence, water from the Warm Deep current rises to within 100 meters of the surface.

Arrow worms Torpedo-shaped members of the plankton, which are voracious feeders on copepods. They belong to the phylum Chaetognatha.

Arthropods The "jointed leg" animals—the largest single phylum on earth, consisting of such familiar forms as insects and crustaceans.

Ascidians Sessile urochordates known as "sea squirts." They attach to rocks and shells on the bottom and filter plankton from sea water. Also known as "tunicates."

Austral Referring or pertaining to the Southern Hemisphere.

Bathydraconidae A family of nototheniiform fishes found in the Southern Ocean, known as the Antarctic dragonfishes.

Bathypelagic Zone The zone of the deep oceans to which absolutely no sunlight penetrates, generally from about 1,000 meters to just above the bottom.

Benthos The region of the ocean bottom which is inhabited by plants and animals that live on, in, or just above sea floor.

Biomass A measure of the amount of life occurring in a circumscribed region. It is often expressed as the weight of all the living material present; either as "wet" weight, or as "dry" weight—sans water.

Bivalves Mollusks of the Class Bivalvia (also known as Pelecypoda or Lamellibranchia), which have shells consisting of two hinged valves. They are familiar as clams, scallops, and oysters.

Boreal Pertaining to the North or the Northern Hemisphere.

Brachiopods Sometimes known as "lamp shells," this phylum of animals only superficially resemble bivalve mollusks. These sessile, often stalked invertebrates, are continental shelf, marine dwellers. Although a few species survive today, more than 30,000 species are known from Paleozoic and Mesozoic marine deposits, where they are often exceedingly abundant.

Brash Ice Small chunks of ice covering the sea surface, usually representing the final stages of the disintegration of icebergs and pack ice.

Brotulidae A family of deep-sea fishes, some of whose members penetrate south into the deeper parts of the Southern Ocean.

Bryophytes The phylum of primitive vascular plants known as mosses.

Calyptopis A stage in the life cycle of krill.

Cambrian Period The earliest geological period in the Paleozoic Era, beginning almost 600 million years ago and lasting for from 70 to 100 million years. Brachiopod and trilobite fossils are known from this period.

Capelin A small schooling fish, *Mallotus villosus,* of the family Osmeridae (smelts), that is abundant in the North Atlantic Ocean.

Caruncles Swollen fleshy nodules at the base of the bill in some birds such as the cormorants.

Channichthyidae The family of nototheniiform fishes that characteristically lack hemoglobin in the blood.

Charadriiformes The order of birds that includes the skuas, gulls, terns, and shorebirds among others.

Chelicerae Pincer appendages of certain crustaceans.

Chionidae The family of birds to which the sheathbills belong.

Chironomidae The family of insects commonly known as midges.

Chrysophyta Phylum of protists known as the "golden algae," including in some classifications the diatoms.

Cladocerans Planktonic crustaceans known as water fleas that are particularly abundant in fresh water ecosystems.

Convection Winds Winds driven by the convective rising of warm air, coupled by the distant sinking of a cold air mass. Air flowing over the earth's surface from the site of sinking to that of rising create these winds.

Copepods Planktonic marine curstaceans that are but a few millimeters in length, and which are often dominant members of the zooplankton community. Many forms are parasites of fish.

Craton The stable central portion of a continental land mass.

Crepuscular Active in dim light, such as at dawn or dusk.

Cretaceous The concluding period of the Mesozoic Era, beginning about 135 million years ago and continuing for 70 million years. The conclusion of the Cretaceous is considered by some to have been precipitous. Although many archaic birds and reptiles became extinct throughout this period, widespread extinctions occurred at its end.

Crinoids The sea lilies are stalked echinoderms that were abundant in Paleozoic seas, although only about 80 species survive—mostly in deep water. They represent one of the most ancient groups of echinoderms, which today are dominated by sea stars, brittle stars, and sea urchins.

Crustose Lichens Lichens with an encrusting growth form, flattened against the rocks on which they develop.

Cryoplankton Those planktonic organisms (principally phytoplankton, but including some zooplankton) that either are frozen into sea ice and grow there, or are closely associated or appended to the underside of the sea ice.

Cryptogamous Descriptive of "lower" plants in an older taxonomy, including small, cryptic forms such as algae, mosses, lichens, blue-green algae, and ferns.

Decapoda This large order (8,500 species) of crustaceans includes forms familiar as shrimp, lobsters, and crabs. Very few decapods are found in the Southern Ocean.

Delphinidae The largest and most diverse family of odontocete cetaceans. They characteristically have pronounced "melons" on the head, and most have the mouth extended into a beak. The Killer whale is the largest delphinid.

Detritivores Animals that graze on partially decomposed organic matter that accumulates in terrestial ecosystems, and on the bottom in aquatic ones.

Diatoms The most important photosynthetic protists found in the Southern Ocean. Belonging to the Chrysophyta, they have silica-impregnated shells (frustules), and can be found in enormous densities, staining the water brown.

Dinoflagellates Photosynthetic protists of the phylum Pyrrophyta that characteristically have two flagellae. Many species are bioluminescent. Although they are exceedingly important in temperate and tropical seas, their presence in the Southern Ocean is overshadowed by diatoms.

Diomedeidae The family of procellariiform birds that includes that includes several species of albatross and mollymawks.

Drift Ice Ice which has broken free from the solid pack ice and which is being further eroded by the action of ocean swells. It usually consists of small to intermediate-sized floes separated by open water.

East Wind Drift A poorly defined surface current system that moves counterclockwise around the Antarctic continent south of about 65° S. It is driven by the prevailing easterlies at these latitudes, and deflects off topographic features of the coast (like the Peninsula) to create gyres along the southern edge of the West Wind Drift.

Elasmobranchs The cartilagenous fishes, including the sharks, skates, and rays.

Eocene This epoch of the Tertiary Period (Cenozoic Era) began approximately 56 million years ago and lasted for about 22 million years. It was during this epoch that the Antarctic Peninsula began to pull away from its association with the South American continent. The Eocene was preceded by the Paleocene and followed by the Oligocene.

Epipelagic Zone The uppermost region of the open ocean's pelagic realm, coincident with the euphotic zone.

Epontic Habitat The life zone created by the presence of sea ice for protists and other plankton that live within the ice, or just beneath its surface.

Eucarida The superorder of crustaceans that contain both euphausiids and decapods. All have stalked compound eyes and fused thoracic segments.

Euphausiids Krill—eucarid marine crustaceans with the gills poorly protected. Most are filter feeders.

Euphotic Zone The zone in an oceanic or freshwater ecosystem through which light can penetrate with enough intensity to provide photosynthetic organisms with their energy requirements. The "sunlit" zone.

Eutrophic Meaning rich in energy. In aquatic ecosystems the chemical richness of eutrophication sometimes creates conditions of anoxia and other undesirable effects.

Falcate Fins Slender fins which curve or arch backward, appearing streamlined, and associated with fast-swimming animals.

Fetch The distance over which wind can maintain an uninterupted contact with water. Generally, the greater the fetch, the larger the waves that will result from a given wind velocity and water depth.

Foliose Lichens Lichens with the thalli raised as thin leaf-like projections.

Foraminifera Protists belonging to the phylum Sarcodina (amoebas), which enclose themselves within a mineral (usually calcium carbonate) shell or test. The foram can secrete additional layers of shell, growing as it does. Some forams reach diameters of several centimeters. While most foraminifera are benthic marine organisms, many planktonic species also occur.

Frazil Ice Flattened, elongate crystals of ice that result from the freezing of sea water. Unconsolidated frazil has a soupy or mushy consistency.

Frustules The shells or tests of diatoms and other phytoplankton. They are typically made of silica-containing compounds.

Fruticose Lichens Lichens with a growth form that features thickened, fleshy thalli that are raised well above the substrate.

Fucoxanthin An accessory photosynthetic pigment typical of the Chrysophyta, including diatoms.

Furcilia A larval stage in the life history of euphausiids.

Glycoproteins A group of proteins that are able to interact with the growing edges of ice crystals, retarding the formation of ice and acting as effective antifreezes.

Grenadiers A family of teleost fish (Macruouridae) that are closely related to the cods and are abundant in deeper boreal waters.

Gyre A limb of an ocean current reflected back onto itself so as to produce a circular pattern.

Heterotrophic Literally meaning a "different source of energy." Pertaining to those organisms that must derive their energy from complex organic molecules. In other words, those that must eat food.

Hummocky Ice Old pack or fast ice that has suffered forces (internal expansion or collisions) that cause it to be thrown up into irregular ridges.

Isopods Benthic crustaceans that are dorso-ventrally flattened and are usually 5 to 15 millimeters in length. Most are scavengers or detritus feeders.

Isostatic Movements Vertical adjustments in the earth's crust that result from changes in its burden.

Jurassic The middle period of the Mesozoic Era, beginning about 180 million years ago and lasting for 45 million years. The earliest birds appeared in this time.

Katabatic Winds Gravity winds created when cold dense air slides down precipitous slopes from the interior ice sheet to the coast. These winds frequently follow periods of radiational cooling. Their severity is determined by topography.

Krill Literally meaning "whale food," but in the Antarctic referring to the most abundant forage for whales, euphausiids of the genus *Euphausia*, particularly *Euphausia superba*.

Lanugo Fur The newborn fur of seals, particularly the ice-breeding species. This is shed and is replaced several weeks after birth by a shorter, coarser adult fur.

Liparidae A family of small, benthic fishes known as "seasnails." They are most abundant in the Northern Hemisphere, but several species are found deep on the Antarctic continental shelf.

Macrouridae The family of teleost fishes known as the grenadiers. Closely related to cods, they are dominant elements of the boreal fish fauna in the deeper parts of the continental shelf.

Mesopelagic Zone The pelagic zone of the open ocean, bounded at its top by the epipelagic zone and at its bottom by the bathypelagic zone.

Mesozoic Era The "Age of Reptiles," beginning with the Triassic Period 225 million years ago, and ending with the close of the Cretaceous 65 million years ago.

Metanauplius A planktonic stage in the larval development of crustaceans.

Miocene An epoch of the Teritary Period of the Cenozoic Era, commencing 25 million years ago and ending 13 million years later.

Mosasaurs A group of marine lizards that were distributed all over the world during the Cretaceous and became extinct at the end of that period.

Muraenolepididae The family of fishes known as eel cods, two species of which are found in the Southern Ocean. These benthic fishes belong to the order (Gadiformes) that include the true codfishes of the Northern Hemisphere.

Mysids Small, shrimp-like marine crustaceans that bear a ventral marsupium and, because of this, are called "opposum shrimp."

Mysticetes The order of cetaceans that include the baleen, or whalebone, whales. All of these filter plankton or small fishes as their source of food.

Nannoplankton Extremely small planktonic organisms, that are difficult to sample with ordinary plankton nets.

Nauplius A planktonic, early stage in the larval development of crustaceans.

Neritic Zone The near-shore, shallow water zone on the continental shelf.

Nototheniiformes An order of teleosts that is the principal group of fishes found in the Southern Ocean.

Nunataks Isolated spires or ridges of rock that project through the ice sheet.

Odontocetes The order of toothed whales that include dolphins, porpoises, and Sperm whales.

Oegopsida A group of oceanic, deep-sea squids that are principal denizens of the Southern Ocean.

Oligocene The third epoch of the Tertiary Period, beginning about 34 million years ago and ending 9 million years later.

Oligotrophic Meaning little energy. Used to describe aquatic ecosystems that are relatively devoid of nutrients, and consequently are poorly productive. Characterized by clear water, populated by relatively few organisms.

Paleocene The earliest epoch of the Tertiary Period, beginning 65 million years ago and ending with the start of the Eocene, 56 million years ago.

Paleomagnetism When molten rock solidifies, the iron-containing minerals become magnitized and are oriented with respect to the earth's magnetic field. By examining this orientation geologists can determine the position of the rock with respect to the polar position at the time the rock formed.

Paleozoic Era This geological era began with the Cambrian Period 600 million years ago and closed with the end of the Permian Period, 225 million years ago. It witnessed the rise of vertebrates, including fishes, amphibians, and reptiles.

Pelecaniformes The order of sea birds that includes gannets, boobies, pelicans, and cormorants.

Pennate Diatoms Diatoms that have a longitudinal axis, along which the two valves of the frustule articulate. Opposed to centric diatoms, which are radially symmetrical.

Peritrophic Membrane A chitenous membrane secreted by the gut around a bolus of food in the digestive tract of copepods and other planktonic crustaceans. The peritrophic membrane is excreted as the binding structure of the fecal pellet.

Phalacrocoracidae The family of birds known as cormorants or shags.

Phenotype The visible expression of an organism's genotype. How it appears.

Phocidae The family of "earless seals" to which belong some of the principal seals of the Antarctic, including the southern elephant seal, the crabeater seal, the leopard seal, and the Weddell seal.

Photophores Bioluminescent skin organs found in some groups of mesopelagic and bathypelagic teleosts such as the lantern fishes.

Pinnipeds The suborder of carnivorous mammals that we know as seals and walruses.

Pleopods The anterior abdominal appendages of crustaceans such as krill. These are often equipped with setae to assist in swimming or filter feeding.

Plesiosaurs Mesozoic marine reptiles that featured long necks and paddle-like limbs. They appeared at the end of the Triassic and became extinct by the end of the Cretaceous.

Polynyas Persistent open leads or pools found deep in the winter pack ice in some regions of the Southern Ocean. Animals such as Killer whales have been observed to over-winter in such situations.

Procellariidae The family of shearwaters and petrels.

Procellariiformes An order of sea birds, including albatross, petrels, shearwaters, and storm petrels; many of which occur in the Southern Ocean.

Protists Single-celled eukaryotic organisms that include several classes of photosynthetic organisms (unicellular algae) and heterotrophs (protozoa).

Proventricular Stomach The glandular, gastric juice secreting portion of the stomach of birds. It is anterior to and distinct from the more muscular part of the stomach—the gizzard.

Pycnogonids A class of marine invertebrates that move slowly over the bottom on long legs. They are known as "sea spiders."

Radiolarans Like foraminifera these are shelled sarcodinian protists that are often abundant in the plankton. They build silicious shells that accumulate in benthic sediments. Fossils of these organisms are known from Precambrian deposits, laid down more than 600 million years ago.

Refugia Protected habitats where elements of a flora and fauna could survive periods of extreme conditions, such as glaciation.

Rorquals Baleen whales of the genus *Balaenoptera,* including the Blue, Fin, and Sei whales.

Rosellidae A family of sponges that are well represented on the Antarctic continental shelf. They have very long silicious spicules that accumulate in the sediments there.

Rotifers A phylum of small freshwater organims that are known from lakes on the Antarctic Peninsula.

Sastrugi Distinctive snow drifts found in the Antarctic that result from sustained unidirectional winds compacting blowing snow. The drifts present an array of irregular lines, with each drift having a flattened crest that frequently overhangs the softer, less well-compacted down wind side.

Setae Fine comb or hair-like structures that often are found on the feeding, respiratory, or locomotor appendages of crustaceans. They seem to be used either to move parcels of water or to filter plankton from it.

Spheniscidae The single family of penguins.

Sphenisciformes The order of birds that are known as penguins.

Spicules Needle-like structures that adorn the surface of animals like sea urchins, or form the skeletal parts of sponges.

Stercorariidae The family of jaegers and skuas—hawklike sea birds with slightly hooked beaks and piratical dispositions.

Subduction The depression of the edge of a tectonic plate that occurs when it is forced under the edge of an adjacent plate. This is thought to be the origin of deep ocean trenches.

Subfossil Remains The remains of dead plants and animals that have not completed the mineralization that produces fossils.

Sublimation The phase transformation from solid to gas without an intervening liquid phase. The "evaporation" of a solid like ice directly to gas (water vapor).

In the Antarctic, sublimation plays an important role in the ablation of the ice sheet.

Tabular Bergs Large flat-topped icebergs that are sometimes many square kilometers in surface extent and more than 100 meters above sea level. They result from the fracture of large blocks off ice shelves.

Talus Angular blocks of rock that have broken off cliffs or steep slopes and accumulate below them.

Tardigrades Tiny (0.3 to 0.5 millimeter) animals known as "water bears" that often live in the film of water found on the "leaves" of mosses and lichens.

Teleosts The bony fishes such as cods, flounders, and salmon. They are distinct from the cartilaginous fishes (sharks and rays), and groups such as sturgeons and gars.

Tertiary The earliest period of the Cenozoic Era, which began about 65 million years ago, after the end of the Cretaceous. The Tertiary is usually divided into five epochs, the Paleocene, Eocene, Oligocene, Miocene, and Pliocene. It ended about 1.5 million years ago.

Tintinnids Protists (ciliates) that have a characteristic cone or bullet shaped test. Although small and relatively unstudied, they are seemingly important elements in the food chain.

Trilobites An extinct group of marine arthropods, which are considered to be the most primitive members of that great group of animals. They were most abundant during the Cambrian and Ordovician periods and disappeared by the end of the Paleozoic.

Trophic Level A conceptual position in the food web of a community or ecosystem. Photosynthesizing organisms at the primary producer trophic level provide the chemical energy that fuels the ecosystem. Herbivores eat the producers and, in turn, are consumed by predators. Each organism, however, usually gets its energy from several sources, not all of which may be from the same trophic level.

Tunicate Sessile or pelagic marine invertebrates sometimes known as sea squirts. They filter sea water across gill slits in their pharyngeal walls, removing plankton as their food.

West Wind Drift The major current of the Southern Ocean, driven clockwise by the prevailing westerlies north of latitude 65° S, and completely circling Antarctica. It is responsible for isolating the continent in a way that produces the extreme climate here. It is the largest current system in the world's ocean.

Zoarcidae A family of benthic fishes known familiarly as eelpouts. These are abundant in boreal waters, but only a few species penetrate the Southern Ocean. They often show viviparous modes of reproduction, or well-developed parental care.

Zooplankton Non-photosynthetic elements of the oceanic plankton that are dominated by crustaceans like copepods, and protists such as foraminifera, tintinnids, and radiolarans.

INDEX

Italicized page numbers refer to illustrations.